The Human Brain during the Second Trimester 96- to 150-mm Crown-Rump Lengths

This eighth of 15 short atlases reimagines the classic 5-volume *Atlas of Human Central Nervous System Development.* This volume presents serial sections from specimens between 96 mm and 150 mm with detailed annotations. An introduction summarizes human CNS developmental highlights between 3.5 and 4.5 months of gestation. The accompanying Glossary (available free online) gives definitions for all the labels used in this volume and all the others in the *Atlas.*

Key Features

- Classic anatomical atlas

- Detailed labeling of structures in the developing brain offers updated terminology and the identification of unique developmental features such as germinal matrices of specific neuronal populations and migratory streams of young neurons

- Appeals to neuroanatomists, developmental biologists, and clinical practitioners

- A valuable reference work on brain development that will be relevant for decades

ATLAS OF
HUMAN CENTRAL NERVOUS SYSTEM DEVELOPMENT
Series

The Human Brain during the Second Trimester 96- to 150-mm Crown-Rump Lengths

Atlas of Human Central Nervous System Development, Volume 8

Shirley A. Bayer
Joseph Altman

CRC Press
Taylor & Francis Group
Boca Raton London New York

CRC Press is an imprint of the
Taylor & Francis Group, an **informa** business

First edition published 2023
by CRC Press
6000 Broken Sound Parkway NW, Suite 300, Boca Raton, FL 33487-2742

and by CRC Press
4 Park Square, Milton Park, Abingdon, Oxon, OX14 4RN

CRC Press is an imprint of Taylor & Francis Group, LLC

LCCN no. 2022008216

ISBN: 978-1-032-22463-3 (hbk)
ISBN: 978-1-032-22461-9 (pbk)
ISBN: 978-1-003-27266-3 (ebk)

DOI: 10.1201/9781003272663

Typeset in Times Roman
by KnowledgeWorks Global Ltd.

Printed in the UK by Severn, Gloucester on responsibly sourced paper

Access the Support Material: www.routledge.com/9781032224633

CONTENTS

ACKNOWLEDGMENTS

We thank the late Dr. William DeMyer, pediatric neurologist at Indiana University Medical Center, for access to his personal library on human CNS development. We also thank the staff of the National Museum of Health and Medicine that were at the Armed Forces Institute of Pathology, Walter Reed Hospital, Washington, D.C. when we collected data in 1995 and 1996: Dr. Adrianne Noe, Director; Archibald J. Fobbs, Curator of the Yakovlev Collection; Elizabeth C. Lockett; and William Discher. We are most grateful to the late Dr. James M. Petras at the Walter Reed Institute of Research who made his darkroom facilities available so that we could develop all the photomicrographs on location rather than in our laboratory in Indiana. Finally, we thank Chuck Crumly, Neha Bhatt, Kara Roberts, Michele Dimont, and Rebecca Condit for expert help during production of the manuscript.

AUTHORS

Shirley A. Bayer received her PhD from Purdue University in 1974 and spent most of her scientific career working with Joseph Altman. She was a professor of biology at Indiana-Purdue University in Indianapolis for several years, where she taught courses in human anatomy and developmental neurobiology while continuing to do research in brain development. Her lengthy publication record of dozens of peer-reviewed, scientific journal articles extends back to the mid 1970s. She has co-authored several books and many articles with her late spouse, Joseph Altman. It was her research (published in *Science* in 1982) that proved that new neurons are added to granule cells in the dentate gyrus during adult life, a unique neuronal population that grows. That paper stimulated interest in the dormant field of adult neurogenesis.

Joseph Altman, now deceased, was born in Hungary and migrated with his family via Germany and Australia to the US. In New York, he became a graduate student in psychology in the laboratory of Hans-Lukas Teuber, earning a PhD in 1959 from New York University. He was a postdoctoral fellow at Columbia University, and later joined the faculty at the Massachusetts Institute of Technology. In 1968, he accepted a position as a professor of biology at Purdue University. During his career, he collaborated closely with Shirley A. Bayer. From the early 1960s-2016, he published many articles in peer-reviewed journals, books, monographs, and free online books that emphasized developmental processes in brain anatomy and function. His most important discovery was adult neurogenesis, the creation of new neurons in the adult brain. This discovery was made in the early 1960s while he was based at MIT, but was largely ignored in favor of the prevailing dogma that neurogenesis is limited to prenatal development. After Dr. Bayer's paper proved new neurons are added to granule cells in the hippocampus, Dr. Altman's monumental discovery became more accepted. During the 1990s, new researchers "rediscovered" and confirmed his original finding. Adult neurogenesis has recently been proven to occur in the dentate gyrus, olfactory bulb, and striatum through the measurement of Carbon-14—the levels of which changed during nuclear bomb testing throughout the 20th century—in postmortem human brains. Today, many laboratories around the world are continuing to study the importance of adult neurogenesis in brain function. In 2011, Dr. Altman was awarded the Prince of Asturias Award, an annual prize given in Spain by the Prince of Asturias Foundation to individuals, entities, or organizations globally who make notable achievements in the sciences, humanities, and public affairs. In 2012, he received the International Prize for Biology - an annual award from the Japan Society for the Promotion of Science (JSPS) for "outstanding contribution to the advancement of research in fundamental biology." This Prize is one of the most prestigious honors a scientist can receive. When Dr. Altman died in 2016, Dr. Bayer continued the work they started over 50 years ago. In her late husband's honor, she created the Altman Prize, awarded each year by JSPS to an outstanding young researcher in developmental neuroscience.

INTRODUCTION

A. Specimens and Organization

Volume 8 in the *Atlas of Human Central Nervous System Development* series illustrates development of the human brain between 3.5 and 4.5 months in the early second trimester, originally presented in Bayer and Altman 2005). The specimens presented here are nearing anatomical maturity throughout the diencephalon, midbrain, pons, and medulla, where few remnants of migrating streams of neurons and the germinal matrices remain. In contrast, the telencephalon (cerebral cortex, basal ganglia) contains many migrating neurons and active embryonic germinal matrices—the primary neuroepithelium and the secondary subventricular and subgranular zones. In addition, immature features predominate in the cerebellum; most of its neurons are still migrating, and its secondary germinal matrix, the external germinal layer, completely covers the surface.

Grayscale photographs of Nissl-stained sections of three normal brains in the Yakovlev Collection[1] are illustrated. **Part II** presents Y144-63 (96-mm crown-rump length) cut in the frontal plane. **Part III** presents Y37-63 (145-mm crown-rump length) cut in the sagittal plane. **Part IV** presents Y15-63 (145-mm crown-rump length) cut in the frontal plane. Frontal plates are presented in serial order from rostral to caudal; the dorsal part of each section is toward the top of the page, the ventral part at the bottom, and the midline is in the vertical center of each section. Sagittal plates are ordered from medial to lateral; the anterior part of each section is facing to the left, posterior to the right. **Part A** of each plate on the left page shows the full-contrast photograph without labels; **Part B** shows low-contrast copies of the same photograph on the right page with superimposed outlines

of structures and unabbreviated labels. The *low-magnification plates* show entire sections to identify large structures of the bra. The *high-magnification plates* feature enlarged views of the brain core to identify smaller structures. A few *very-high-magnification plates* show the cerebral cortex in great detail. As in other volumes, transient structures are labeled in ***italics***. During fixation, shrinkage introduced artifactual infolding of the cerebral cortex in some specimens. During dissection, embedding, cutting, and staining, some of the sections illustrated were torn. Both artifacts and processing damage are usually outlined with *dashed lines* in part **B** of each plate.

B. Developmental Highlights

A side-by-side comparison of younger and older frontally sectioned brains at anterior (**Figs. 1-2**), middle (**Figs. 3-4**), and posterior (**Figs. 5-6**) levels of the telencephalon will highlight major developmental changes in the early second trimester. Illustrations are modified from our book on the *Development of the Human Neocortex* (Altman and Bayer, 2015) that is available free online at neurondevelopment.org.

---

1. The *Yakovlev Collection* (designated by a **Y** prefix in the specimen number) is the work of Dr. Paul Ivan Yakovlev (1894–1983), a neurologist affiliated with Harvard University and the AFIP. Throughout his career, Yakovlev collected many diseased and normal human brains. He invented a giant microtome that was capable of sectioning entire human brains. Later, he became interested in the developing brain and collected many human brains during the second and third trimesters. The normal brains in the developmental group were cataloged by Haleem (1990) and were examined by us during 1996 and 1997. The collection was moved to the National Museum of Health and Medicine when the Armed Forces Institute of Pathology (AFIP) closed and is still available for research.

REFERENCES

Altman J, Bayer SA. (2015) *Development of the human neocortex*. Ocala, FL, Laboratory of Developmental Neurobiology, neurondevelopment.org.

Bayer SA, Altman J (2005) *Atlas of Human Central Nervous System Development*, Volume 3: *The Human Brain during the Second Trimester*. Boca Raton, FL, CRC Press.

Bayer SA, Altman J (in press) *The Human Brain during the First Trimester 57- to 60-mm Crown-Rump Lengths, Atlas of Human Central Nervous System Development*, Volume 7, CRC Press/Taylor and Francis.

Haleem M (1990) *Diagnostic Categories of the Yakovlev Collection of Normal and Pathological Anatomy and Development of the Brain*. Washington, D.C. Armed Forces Institute of Pathology.

Retzius, G. (1896) *Das Menschenhirn: Studien in der makroskopischen Morphologie*. Vols. 1 and 2. Stockholm: Königliche Buchdruckerei.

THE BRAIN OF A 3.5-MONTH FETUS-ANTERIOR TELENCEPHALON
GW14

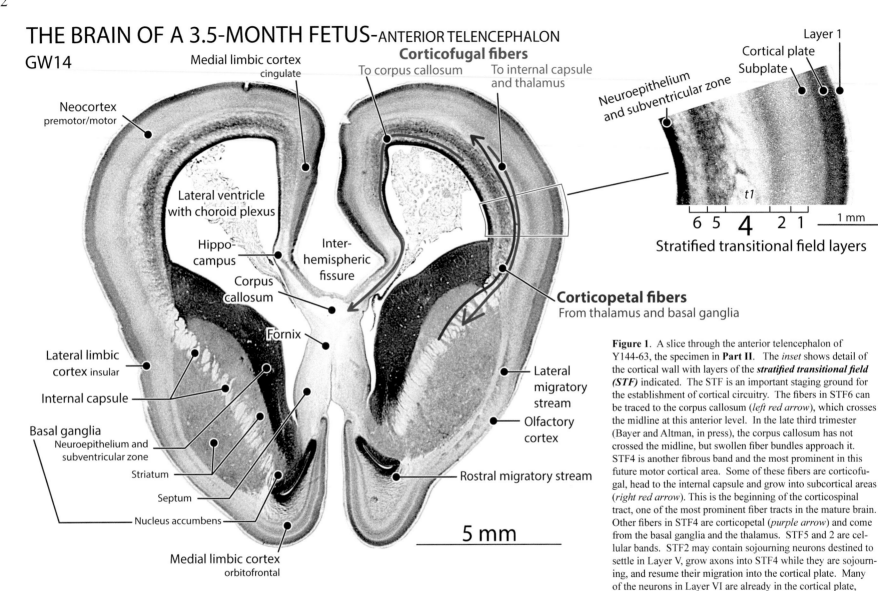

Medial limbic cortex cingulate

Corticofugal fibers
To corpus callosum   To internal capsule and thalamus

Layer 1
Cortical plate
Subplate
Neuroepithelium and subventricular zone

Neocortex premotor/motor

Lateral ventricle with choroid plexus

Hippo-campus

Inter-hemispheric fissure

Corpus callosum

Fornix

Lateral limbic cortex insular

Internal capsule

Basal ganglia
Neuroepithelium and subventricular zone
Striatum
Septum
Nucleus accumbens

Medial limbic cortex orbitofrontal

Corticopetal fibers
From thalamus and basal ganglia

Lateral migratory stream

Olfactory cortex

Rostral migratory stream

t1

6 5 **4** 2 1   1 mm

Stratified transitional field layers

5 mm

Figure 1. A slice through the anterior telencephalon of Y144-63, the specimen in **Part II**. The *inset* shows detail of the cortical wall with layers of the ***stratified transitional field (STF)*** indicated. The STF is an important staging ground for the establishment of cortical circuitry. The fibers in STF6 can be traced to the corpus callosum (*left red arrow*), which crosses the midline at this anterior level. In the late third trimester (Bayer and Altman, in press), the corpus callosum has not crossed the midline, but swollen fiber bundles approach it. STF4 is another fibrous band and the most prominent in this future motor cortical area. Some of these fibers are corticofugal, head to the internal capsule and grow into subcortical areas (*right red arrow*). This is the beginning of the corticospinal tract, one of the most prominent fiber tracts in the mature brain. Other fibers in STF4 are corticopetal (*purple arrow*) and come from the basal ganglia and the thalamus. STF5 and 2 are cellular bands. STF2 may contain sojourning neurons destined to settle in Layer V, grow axons into STF4 while they are sojourning, and resume their migration into the cortical plate. Many of the neurons in Layer VI are already in the cortical plate, having migrated the short distance through STF1, through the subplate, to settle in the cortical plate. STF5 is a sojourn zone for recently generated neurons. At this time, they are most likely neurons that will settle in Layers IV and III. STF 5 is not prominent because sensory information is less robust here. (Modified Figure 19A in Altman and Bayer, 2015.)

THE BRAIN OF A 4.5-MONTH FETUS-ANTERIOR TELENCEPHALON
GW17

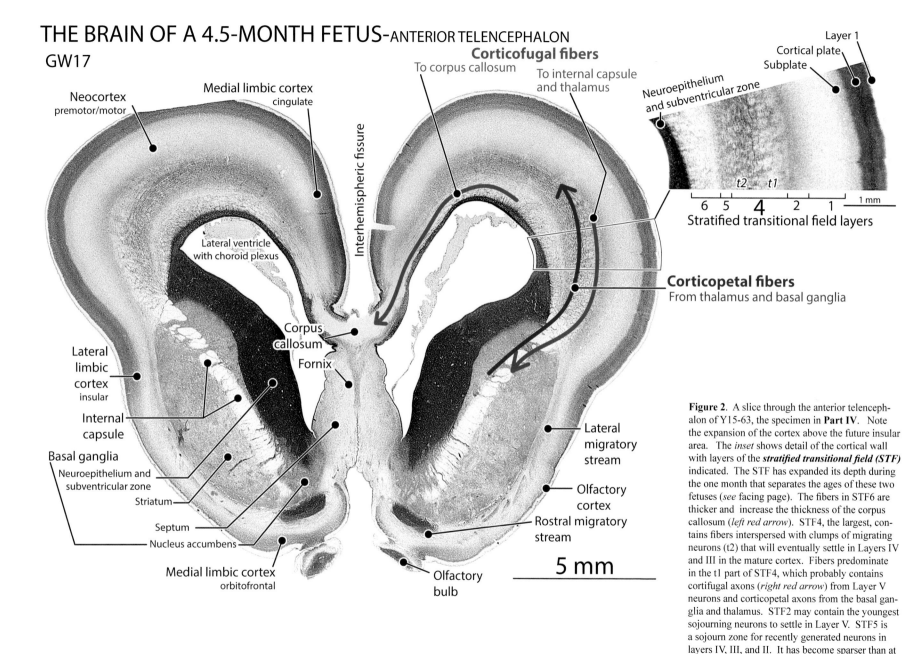

Neocortex
premotor/motor

Medial limbic cortex
cingulate

Corticofugal fibers
To corpus callosum

To internal capsule
and thalamus

Interhemispheric fissure

Lateral ventricle
with choroid plexus

Corpus
callosum

Fornix

Lateral
limbic
cortex
insular

Internal
capsule

Basal ganglia

Neuroepithelium and
subventricular zone

Striatum

Septum

Nucleus accumbens

Medial limbic cortex
orbitofrontal

Olfactory
bulb

Lateral
migratory
stream

Olfactory
cortex

Rostral migratory
stream

5 mm

Layer 1
Cortical plate
Subplate

Neuroepithelium
and subventricular zone

t2   t1

6 5 **4** 2 1   1 mm

Stratified transitional field layers

Corticopetal fibers
From thalamus and basal ganglia

Figure 2. A slice through the anterior telencephalon of Y15-63, the specimen in **Part IV**. Note the expansion of the cortex above the future insular area. The *inset* shows detail of the cortical wall with layers of the ***stratified transitional field (STF)*** indicated. The STF has expanded its depth during the one month that separates the ages of these two fetuses (*see* facing page). The fibers in STF6 are thicker and increase the thickness of the corpus callosum (*left red arrow*). STF4, the largest, contains fibers interspersed with clumps of migrating neurons (t2) that will eventually settle in Layers IV and III in the mature cortex. Fibers predominate in the t1 part of STF4, which probably contains cortifugal axons (*right red arrow*) from Layer V neurons and corticopetal axons from the basal ganglia and thalamus. STF2 may contain the youngest sojourning neurons to settle in Layer V. STF5 is a sojourn zone for recently generated neurons in layers IV, III, and II. It has become sparser than at 3.5 months (facing page). (Modified Figure 19B in Altman and Bayer, 2015.)

4

GW14

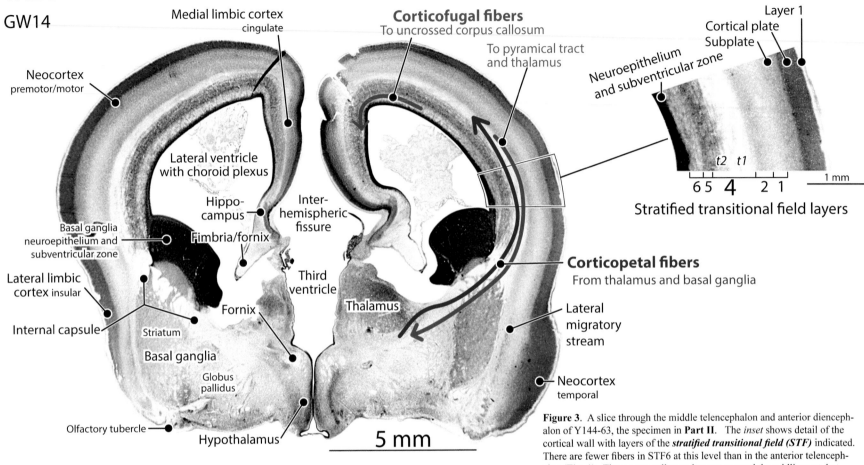

Corticofugal fibers
To uncrossed corpus callosum

To pyramical tract
and thalamus

Medial limbic cortex
cingulate

Neocortex
premotor/motor

Lateral ventricle
with choroid plexus

Hippo-
campus

Inter-
hemispheric
fissure

Fimbria/fornix

Basal ganglia
neuroepithelium and
subventricular zone

Lateral limbic
cortex insular

Internal capsule

Striatum

Fornix

Basal ganglia

Globus
pallidus

Third
ventricle

Thalamus

Olfactory tubercle

Hypothalamus

5 mm

Neuroepithelium
and subventricular zone

Layer 1
Cortical plate
Subplate

t2 t1

6 5 4   2 1

1 mm

Stratified transitional field layers

Corticopetal fibers
From thalamus and basal ganglia

Lateral
migratory
stream

Neocortex
temporal

Figure 3. A slice through the middle telencephalon and anterior diencephalon of Y144-63, the specimen in **Part II**. The *inset* shows detail of the
cortical wall with layers of the *stratified transitional field (STF)* indicated.
There are fewer fibers in STF6 at this level than in the anterior telencephalon (**Fig. 1**). The corpus callosum has not crossed the midline yet, but
fibers in STF6 are approaching it (*left red arrow*). STF4 is a prominent
fibrous band with a superficial part (*t1*), where fibers predominate, and a
deep part (*t2*) where fibers mingle with neurons migrating to the cortical
plate. STF4 fibers are thalamocortical afferents (*purple arrow*) as well
as corticofugal fibers (*right red arrow*). STF5 and 2 are cellular bands.
STF2 is the sojourn zone for Layer V that may be growing axons into
STF4 before resuming their migration into the cortical plate. Most Layer
VI neurons are already in the cortical plate and probably contribute axons
to STF4 that will terminate in the thalamus. STF5 is a sojourn zone for
recently generated neurons. At this time, they are most likely neurons that
will settle in Layers IV, III, and II. This motor/premotor cortical area is
agranular cortex because Layer IV is not prominent in the mature brain.
(Modified Figure 19C in Altman and Bayer, 2015.)

THE BRAIN OF A 4.5-MONTH FETUS-TELENCEPHALON AND DIENCEPHALON
GW17

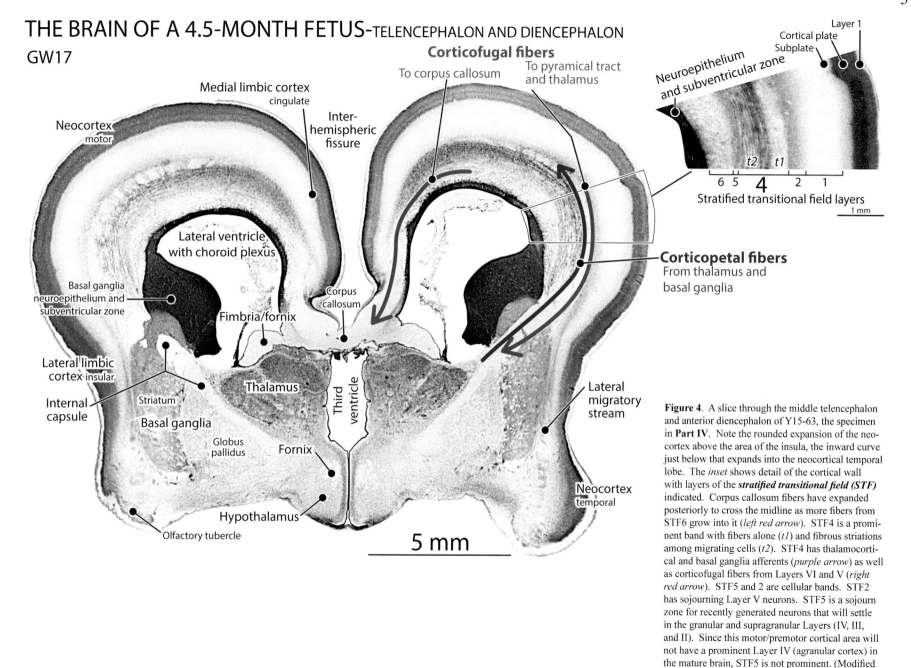

Corticofugal fibers
To corpus callosum · To pyramical tract and thalamus

Neuroepithelium and subventricular zone

Layer 1
Cortical plate
Subplate

t2 *t1*

6 5 **4** 2 1
Stratified transitional field layers

1 mm

Medial limbic cortex
cingulate

Inter-hemispheric fissure

Neocortex
motor

Lateral ventricle
with choroid plexus

Corticopetal fibers
From thalamus and basal ganglia

Basal ganglia
neuroepithelium and
subventricular zone

Corpus callosum

Fimbria/fornix

Lateral limbic
cortex insular

Internal
capsule

Striatum

Thalamus

Third ventricle

Lateral
migratory
stream

Basal ganglia

Globus
pallidus

Fornix

Neocortex
temporal

Hypothalamus

Olfactory tubercle

5 mm

Figure 4. A slice through the middle telencephalon and anterior diencephalon of Y15-63, the specimen in **Part IV**. Note the rounded expansion of the neocortex above the area of the insula, the inward curve just below that expands into the neocortical temporal lobe. The *inset* shows detail of the cortical wall with layers of the ***stratified transitional field (STF)*** indicated. Corpus callosum fibers have expanded posteriorly to cross the midline as more fibers from STF6 grow into it (*left red arrow*). STF4 is a prominent band with fibers alone (*t1*) and fibrous striations among migrating cells (*t2*). STF4 has thalamocortical and basal ganglia afferents (*purple arrow*) as well as corticofugal fibers from Layers VI and V (*right red arrow*). STF5 and 2 are cellular bands. STF2 has sojourning Layer V neurons. STF5 is a sojourn zone for recently generated neurons that will settle in the granular and supragranular Layers (IV, III, and II). Since this motor/premotor cortical area will not have a prominent Layer IV (agranular cortex) in the mature brain, STF5 is not prominent. (Modified Figure 19C in Altman and Bayer, 2015.)

6

THE BRAIN OF A 3.5-MONTH FETUS-
POSTERIOR TELENCEPHALON, MIDBRAIN, AND PONS

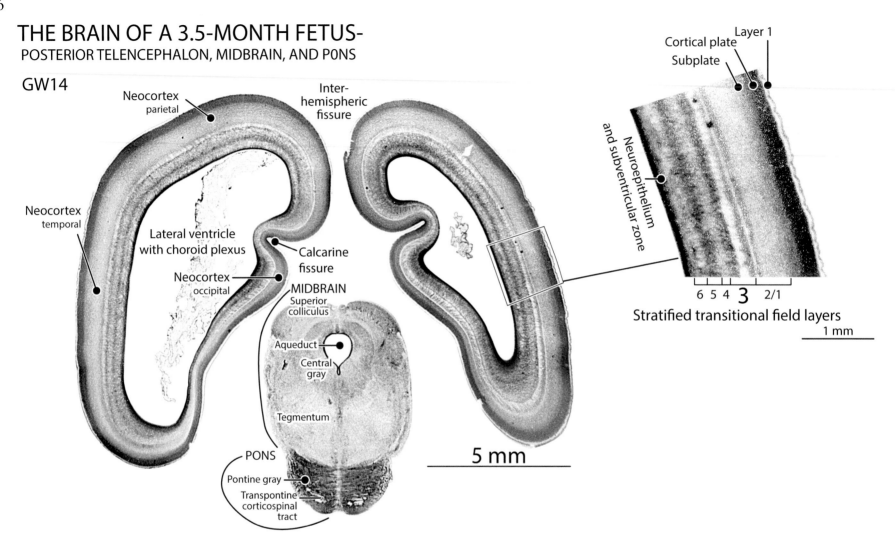

GW14

Neocortex *parietal*

Inter-hemispheric fissure

Neocortex *temporal*

Lateral ventricle with choroid plexus

Neocortex *occipital*

Calcarine fissure

MIDBRAIN
Superior colliculus

Aqueduct

Central gray

Tegmentum

PONS
Pontine gray
Transpontine corticospinal tract

Cortical plate
Layer 1
Subplate

Neuroepithelium and subventricular zone

6 5 4 3 2/1

Stratified transitional field layers

1 mm

5 mm

Figure 5. A slice through the posterior telencephalon, midbrain, and pons of Y144-63, the specimen in **Part II**. Note the calcarine fissure in the medial telencephalon. The *inset* shows detail of the cortical wall with layers of the ***stratified transitional field (STF)*** indicated. This area is sensory cortex where Layer IV is prominent in the mature brain (granular). Consequently, ***STF4*** is much reduced since sensory cortical areas do not project strongly to subcortical structures (*compare* with **Figs. 1** and **3**). ***STF3*** is the most prominent layer here, subdivided into a, b, and c sublayers. Complex interactions between sojourning Layer IV neurons and thalamocortical axons occur in these three sublayers. There are fewer fibers in ***STF6*** at this level than in the anterior and middle telencephalon (**Figs. 1 and 3**) because posterior neocortical areas lag behind anterior areas in maturation. ***STF5*** and ***2*** are cellular bands. ***STF2*** is very thin because Layer V is not prominent in sensory cortical areas. ***STF5*** is a prominent sojourn zone for recently generated neurons in Layers IV to II. At this time, they are most likely Layer IV neurons being prepared for further specification in ***STF3***, and some Layer III neurons.

In the pons, note the very small bundles of axons in the transpontine corticofugal tract; *compare* with **Figure 6** on the facing page. (Modified Figure 19E in Altman and Bayer, 2015.)

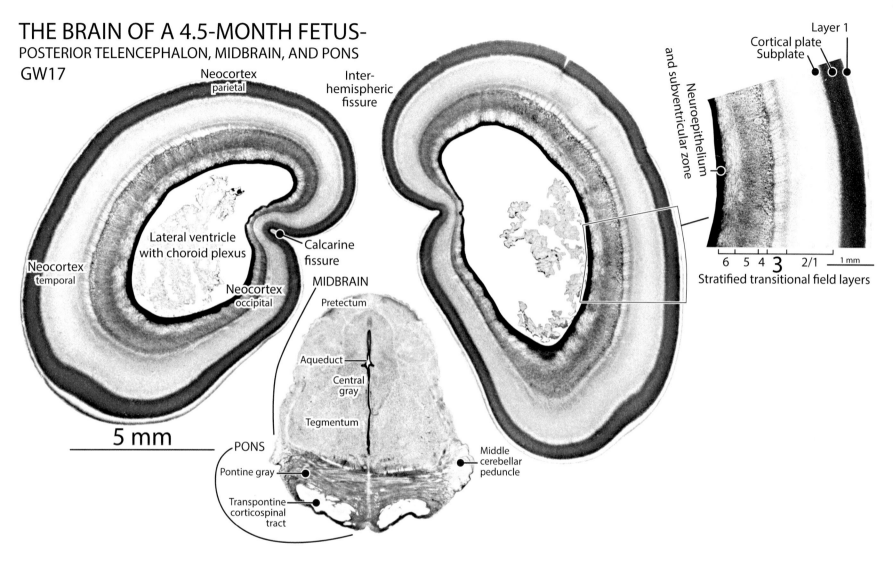

THE BRAIN OF A 4.5-MONTH FETUS-
POSTERIOR TELENCEPHALON, MIDBRAIN, AND PONS
GW17

Figure 6. A slice through the posterior telencephalon, midbrain, and pons of Y15-63, the specimen in **Part IV**. The calcarine fissure is the only notable structure in the medial telencephalon, but the cortex is greatly expanding (*compare* with **Figure 5**). The *inset* shows detail of the cortical wall with layers of the **stratified transitional field (STF)** indicated. Just as in **Figure 5**, **STF4** is much less prominent and **STF2** is nearly absent. The reduction in these two layers is related to the lack of a prominent Layer V in sensory cortical areas. **STF3** is the most prominent layer where interactions with incoming axons from the thalamus contribute to the specification of the multitude of neurons that will occupy Layer IV. Note the expansion of **STF5** compared to that in **Figure 5**. There are more fibers in **STF6** than in the younger fetus on the facing page, and they form clumps rather than a continuous layer.

In the pons, note the much larger bundles of axons in the transpontine corticofugal tract. (Modified Figure 19F in Altman and Bayer, 2015.)

PART II: Y144-63
CR 96 mm (GW 14)
Frontal

This specimen is a stillborn male fetus with a crown-rump length (CR) of 96 mm estimated to be at gestational week (GW) 14 (Yakovlev case number RPSL B-144-63, referred to as Y144-63). The brain was cut in the coronal (frontal) plane in 35-μm thick sections and is classified as a Normative Control in the Yakovlev Collection (Haleem, 1990). Since there is no photograph of this brain before it was embedded and cut, we turned to comprehensive treatise that Retzius published in 1896 showing whole fetal brains in medial, lateral, superior, and inferior views and midline sagittally cut brains. **Figure 7**, taken from Retzius (1896), shows the exterior of a brain from a specimen that is comparable in age to Y144-63, along with the approximate cutting angle of the sections.

Photographs of 16 entire Nissl-stained sections are illustrated in **Plates 1–16. Plates 17 to 24** show high-magnification views of the diencephalon and different areas of the cerebral cortex.

The *cortical neuroepithelium and subventricular zone* is thicker, and the cortical plate is thinner than in the older second-trimester specimens. The *stratified transitional fields* in all lobes of the cerebral cortex are filled with migrating and sojourning neurons and show distinct regional heterogeneity. The cerebral cortex is smooth except for the calcarine sulcus; there is only a slight hint of the lateral sulcus. (The narrow invagina-tions seen in various parts of the cortex are the shrink-age artifacts produced during fixation.) The corpus callosum is only crossed at its anterior part; most callosal fibers are uncrossed. The olfactory bulb is beneath the posterior basal telencephalon and contains the *rostral migratory stream* in its core. In anterolateral parts of the cerebral cortex, streams of neurons and glia abound in the *lateral migratory stream*. Most of the hippocampus is in an immature position dorsal to the thalamus and medial to the temporal lobe. However, there is only a thin *glioepithelium/ependyma* that blends with the *fornical glioepithelium* at the ventricle. Cells are entering Ammon's horn pyramidal layer via a more definite *ammonic migration*, and granule cells and their precursors are migrating to the hilus of the presumptive dentate gyrus in the *dentate migration*; there is no granular layer. A massive *neuroepithelium/subventricular zone* overlies the nucleus accumbens and striatum where neurons (and glia) are being generated. The *striatal neuroepithelium and subventricular zone* have indistinct subdivisions. The *strionuclear glioepithelium* forms definite continuities with other glioepithelia in the telencephalon and diencephalon. The septum has a *glioepithelium/ependyma* at the ventricular surface and its nuclei are generally well-defined.

Neurons in most diencephalic structures appear to be settled; the major exceptions are the immature appearance of the lateral and medial geniculate bodies in the posterior thalamus and the hypothalamic medial mammillary body. The third ventricle is lined by a thin *glioepithelium/ependyma* that shows small invaginations and evaginations that mark previous sites where fate-restricted neuroepithelial patches generated neurons for specific nuclei. In the midbrain, pons, and medulla, there is a *glioepithelium/ependyma* lining the cerebral aqueduct and fourth ventricle, also showing invaginations and evaginations. In the medulla a large *precerebellar neuroepithelium* forms the *ventral lip of the rhombencephalon* and is still generating the youngest pontine gray neurons that are migrating to the pons in the large *anterior extramural migratory stream*. The transpontine corticofugal tract forms a very small, discrete bundle at the base of the pontine gray, and the pyramids are extremely small at the base of the medulla; that indicates that many corticofugal axons have not yet grown into these fiber tracts.

The cerebellum is little more than a thick plate overlying the pons and medulla. The deep nuclei are in place beneath the cortex. The cortical surface is completely covered by an *external germinal layer (egl)*. Lamination in the cortex is nearly absent, except for a thin molecular layer beneath the *egl*. Nearly all Purkinje cells are migrating. Lobulation has barely begun in the vermis and has not yet begun in the hemispheres. The *germinal trigone* is only visible beneath the future flocculus.

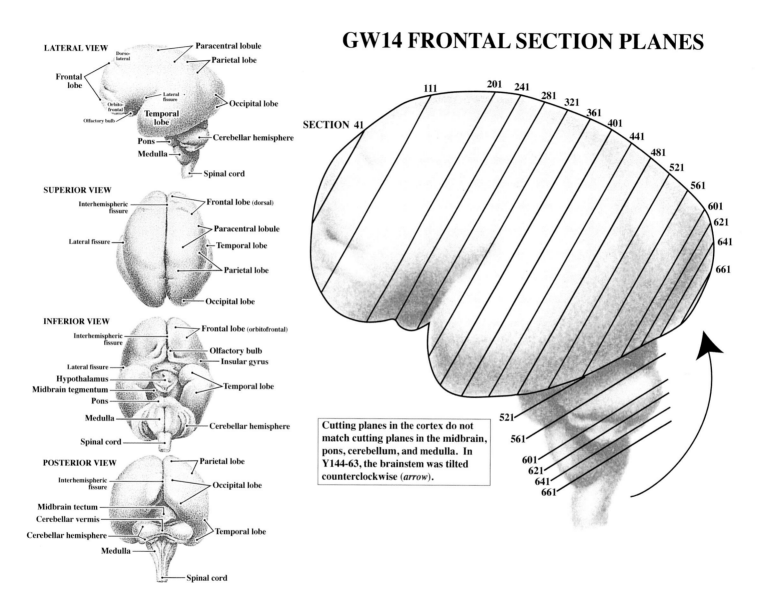

GW14 FRONTAL SECTION PLANES

LATERAL VIEW

Paracentral lobule
Parietal lobe
Dorso-lateral
Frontal lobe
Lateral fissure
Orbito-frontal
Occipital lobe
Temporal lobe
Olfactory bulb
Pons
Cerebellar hemisphere
Medulla
Spinal cord

SUPERIOR VIEW

Interhemispheric fissure
Frontal lobe (dorsal)
Paracentral lobule
Lateral fissure
Temporal lobe
Parietal lobe
Occipital lobe

INFERIOR VIEW

Interhemispheric fissure
Frontal lobe (orbitofrontal)
Olfactory bulb
Insular gyrus
Lateral fissure
Hypothalamus
Midbrain tegmentum
Temporal lobe
Pons
Medulla
Cerebellar hemisphere
Spinal cord

POSTERIOR VIEW

Parietal lobe
Interhemispheric fissure
Occipital lobe
Midbrain tectum
Cerebellar vermis
Cerebellar hemisphere
Temporal lobe
Medulla
Spinal cord

SECTION 41

111 201 241 281 321 361 401 441 481 521 561 601 621 641 661

521 561 601 621 641 661

Cutting planes in the cortex do not match cutting planes in the midbrain, pons, cerebellum, and medulla. In Y144-63, the brainstem was tilted counterclockwise (*arrow*).

Figure 7. The column on the left shows a GW14 brain from lateral, superior, inferior, and posterior views with major structures labeled (Figures 36 to 39 in Table 1, Volume 2, Retzius, 1896). An enlarged lateral view with the approximate locations and cutting angles of the sections of Y144-63 is shown above; the specimen shown above has a more prominent lateral fissure than is found in Y144-63. Note that the cutting angles in the brainstem do not match cutting angles in the cortex. Once the brain has been dissected from the skull, the brainstem has no support to maintain a constant angle of downward extension from the cortex. Consequently, the brainstem can easily be flexed either forward or backward during fixation. In Y144-63, the brainstem was flexed backward and upward (*arrow*) to tuck under the cerebral hemispheres.

PLATE 1A
CR 96 mm, GW 14, Y144-63
Frontal
Section 41

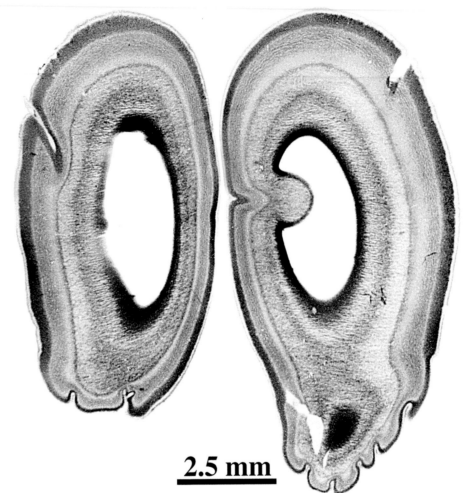

2.5 mm

LAYERS OF THE CORTICAL
STRATIFIED TRANSITIONAL
FIELD (STF)

STF1 Superficial fibrous layer with
an early developmental stage
(t1) when many cells are
migrating through it, followed
by a late stage *(t2)* with sparse
cells. Endures as the
subcortical white matter.

STF2 Upper cellular layer, the last
sojourn zone before cells
translocate to the cortical
plate.

STF4 Complex middle layer with
three developmental stages:
t1– fibrous layer without
interspersed cells; *t2*– cells
and fibers intermingle to form
striations; *t3*– fibers endure in
the deep white matter.

STF5 Deep cellular layer, the first
sojourn zone to appear
outside the germinal matrix.

STF6 Late-forming deep layer of
callosal fibers outside the
germinal matrix.

PLATE 1B

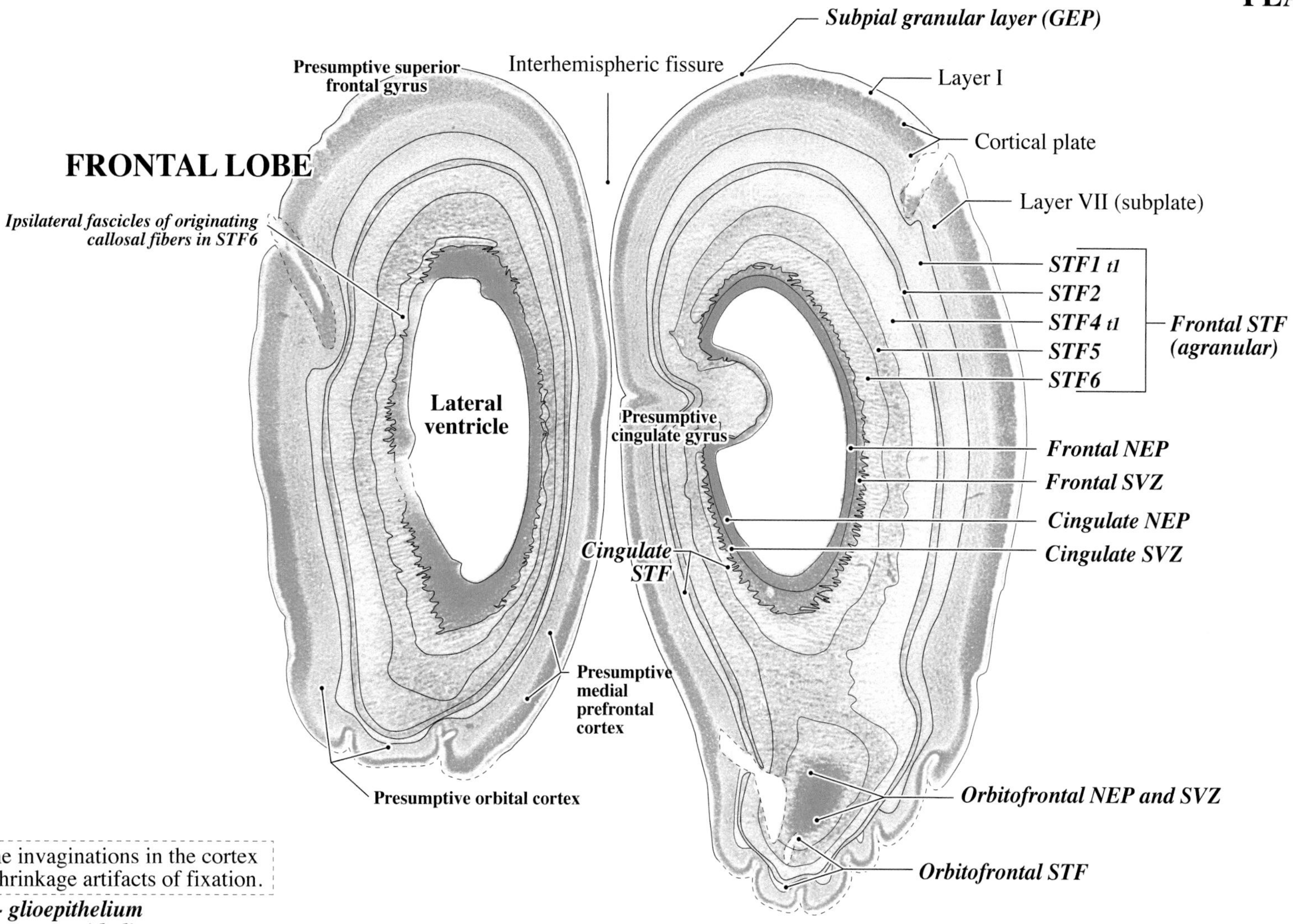

Subpial granular layer (GEP)

Interhemispheric fissure

Layer I

Presumptive superior frontal gyrus

Cortical plate

FRONTAL LOBE

Layer VII (subplate)

Ipsilateral fascicles of originating callosal fibers in STF6

STF1 t1
STF2
STF4 t1
STF5
STF6

Frontal STF (agranular)

Lateral ventricle

Presumptive cingulate gyrus

Frontal NEP
Frontal SVZ

Cingulate NEP
Cingulate SVZ

Cingulate STF

Presumptive medial prefrontal cortex

Presumptive orbital cortex

Orbitofrontal NEP and SVZ

Orbitofrontal STF

Some invaginations in the cortex are shrinkage artifacts of fixation.

GEP - glioepithelium
NEP - neuroepithelium
SVZ - subventricular zone
Germinal and transitional structures in *italics*

PLATE 2A
CR 96 mm, GW 14
Y144-63
Frontal
Section 111

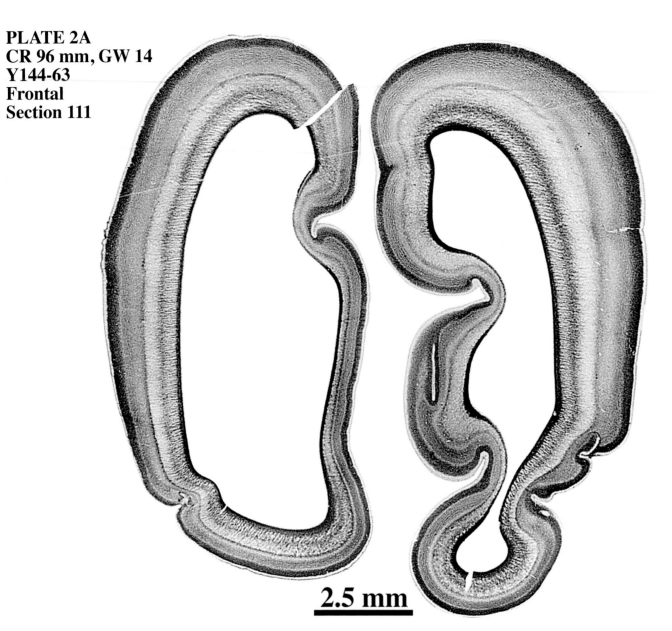

2.5 mm

LAYERS OF THE CORTICAL
STRATIFIED TRANSITIONAL
FIELD (STF)

STF1 Superficial fibrous
layer with an early
developmental stage
(t1) when many cells
are migrating through
it, followed by a late
stage *(t2)* with sparse
cells. Endures as the
subcortical white
matter.

STF2 Upper cellular layer,
the last sojourn zone
before cells
translocate to the
cortical plate.

STF4 Complex middle layer
with three
developmental stages:
t1– fibrous layer
without interspersed
cells; *t2*– cells and
fibers intermingle to
form striations; *t3*–
fibers endure in the
deep white matter.

STF5 Deep cellular layer,
the first sojourn zone
to appear outside the
germinal matrix.

STF6 Late-forming deep
layer of callosal fibers
outside the germinal
matrix.

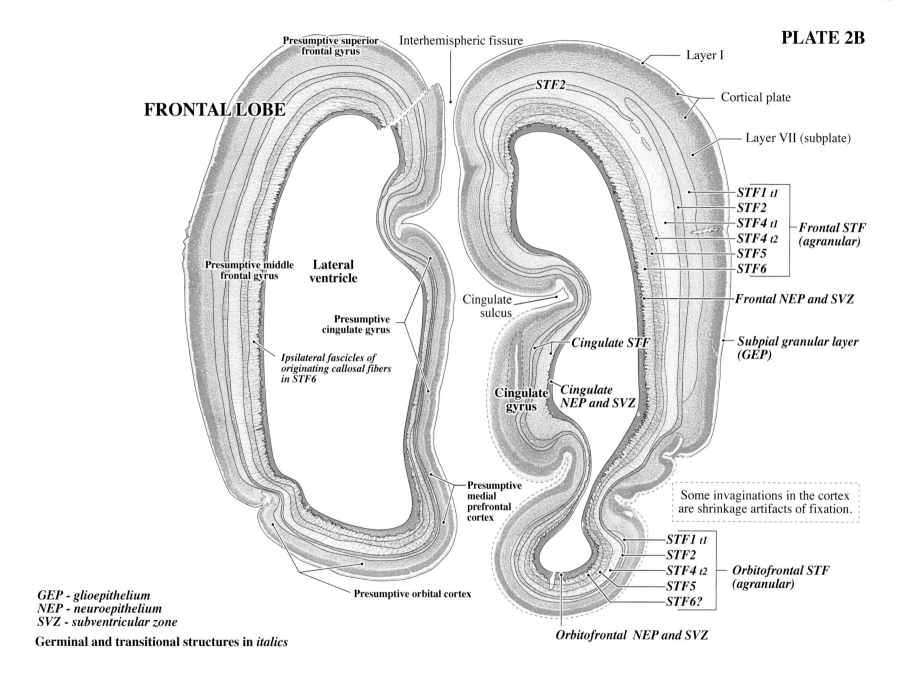

Presumptive superior frontal gyrus

Interhemispheric fissure

PLATE 2B

STF2

Layer I

Cortical plate

FRONTAL LOBE

Layer VII (subplate)

STF1 t1
STF2
STF4 t1
STF4 t2
STF5
STF6

} *Frontal STF (agranular)*

Presumptive middle frontal gyrus

Lateral ventricle

Frontal NEP and SVZ

Cingulate sulcus

Presumptive cingulate gyrus

Cingulate STF

Subpial granular layer (GEP)

Ipsilateral fascicles of originating callosal fibers in STF6

Cingulate gyrus

Cingulate NEP and SVZ

Presumptive medial prefrontal cortex

Some invaginations in the cortex are shrinkage artifacts of fixation.

STF1 t1
STF2
STF4 t2
STF5
STF6?

} *Orbitofrontal STF (agranular)*

Presumptive orbital cortex

GEP - glioepithelium
NEP - neuroepithelium
SVZ - subventricular zone

Germinal and transitional structures in *italics*

Orbitofrontal NEP and SVZ

PLATE 3A
CR 96 mm, GW 14
Y144-63
Frontal
Section 201

**See this area of cortex
in Plate 21.**

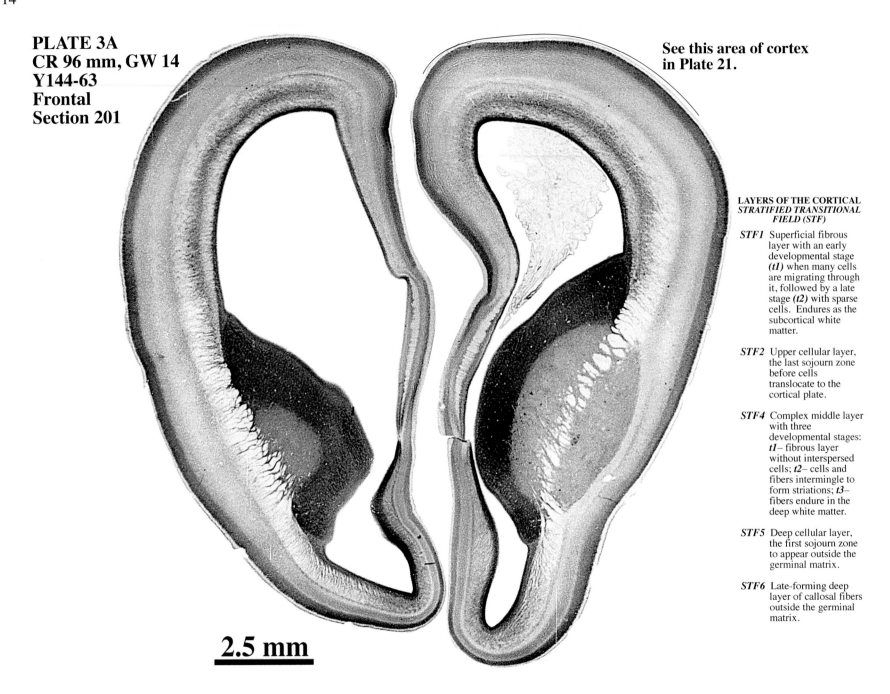

2.5 mm

**LAYERS OF THE CORTICAL
*STRATIFIED TRANSITIONAL
FIELD (STF)***

STF1 Superficial fibrous
layer with an early
developmental stage
(t1) when many cells
are migrating through
it, followed by a late
stage *(t2)* with sparse
cells. Endures as the
subcortical white
matter.

STF2 Upper cellular layer,
the last sojourn zone
before cells
translocate to the
cortical plate.

STF4 Complex middle layer
with three
developmental stages:
t1– fibrous layer
without interspersed
cells; *t2*– cells and
fibers intermingle to
form striations; *t3*–
fibers endure in the
deep white matter.

STF5 Deep cellular layer,
the first sojourn zone
to appear outside the
germinal matrix.

STF6 Late-forming deep
layer of callosal fibers
outside the germinal
matrix.

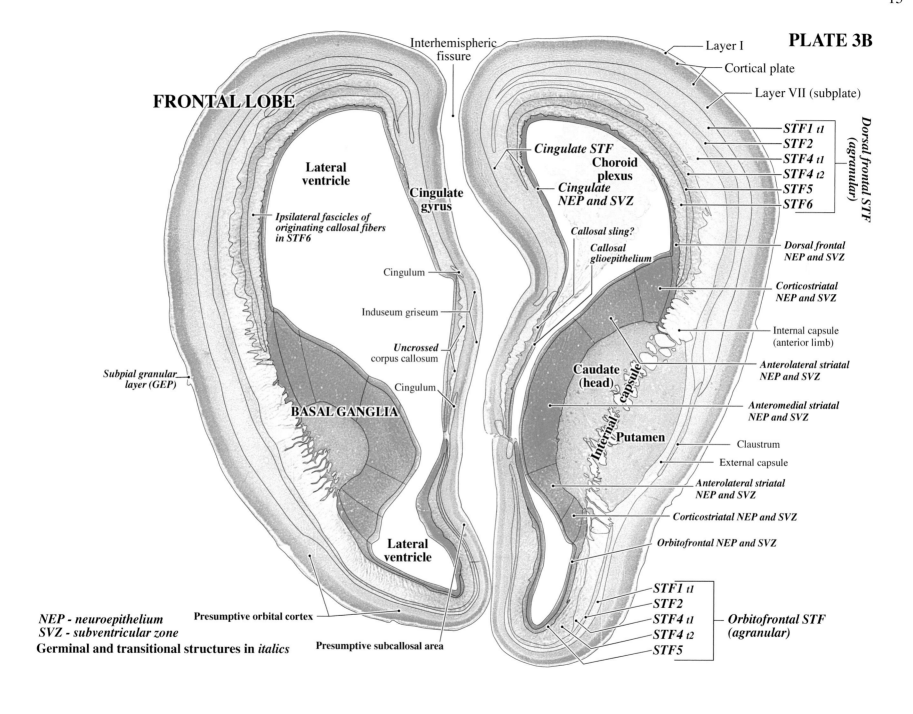

15

PLATE 3B

Interhemispheric fissure

Layer I

Cortical plate

Layer VII (subplate)

FRONTAL LOBE

STF1 t1
STF2
STF4 t1
STF4 t2
STF5
STF6

Dorsal frontal STF (agranular)

Cingulate STF

Choroid plexus

Cingulate NEP and SVZ

Lateral ventricle

Cingulate gyrus

Ipsilateral fascicles of originating callosal fibers in STF6

Callosal sling?

Callosal glioepithelium

Dorsal frontal NEP and SVZ

Cingulum

Corticostriatal NEP and SVZ

Induseum griseum

Internal capsule (anterior limb)

Caudate (head)

Anterolateral striatal NEP and SVZ

Uncrossed corpus callosum

Internal capsule

Cingulum

Anteromedial striatal NEP and SVZ

Subpial granular layer (GEP)

BASAL GANGLIA

Putamen

Claustrum

External capsule

Anterolateral striatal NEP and SVZ

Lateral ventricle

Corticostriatal NEP and SVZ

Orbitofrontal NEP and SVZ

NEP - neuroepithelium
SVZ - subventricular zone
Germinal and transitional structures in *italics*

Presumptive orbital cortex

Presumptive subcallosal area

STF1 t1
STF2
STF4 t1
STF4 t2
STF5

Orbitofrontal STF (agranular)

PLATE 4A
CR 96 mm, GW 14
Y144-63
Frontal
Section 241

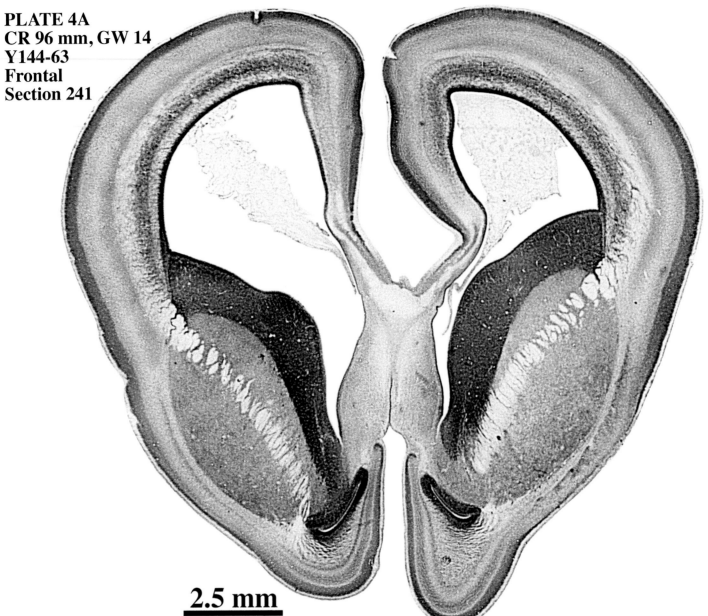

2.5 mm

LAYERS OF THE CORTICAL
STRATIFIED TRANSITIONAL
FIELD (STF)

STF1 Superficial fibrous
layer with an early
developmental stage
(t1) when many cells
are migrating through
it, followed by a late
stage *(t2)* with sparse
cells. Endures as the
subcortical white
matter.

STF2 Upper cellular layer,
the last sojourn zone
before cells
translocate to the
cortical plate.

STF4 Complex middle layer
with three
developmental stages:
t1– fibrous layer
without interspersed
cells; *t2*– cells and
fibers intermingle to
form striations; *t3*–
fibers endure in the
deep white matter.

STF5 Deep cellular layer,
the first sojourn zone
to appear outside the
germinal matrix.

STF6 Late-forming deep
layer of callosal fibers
outside the germinal
matrix.

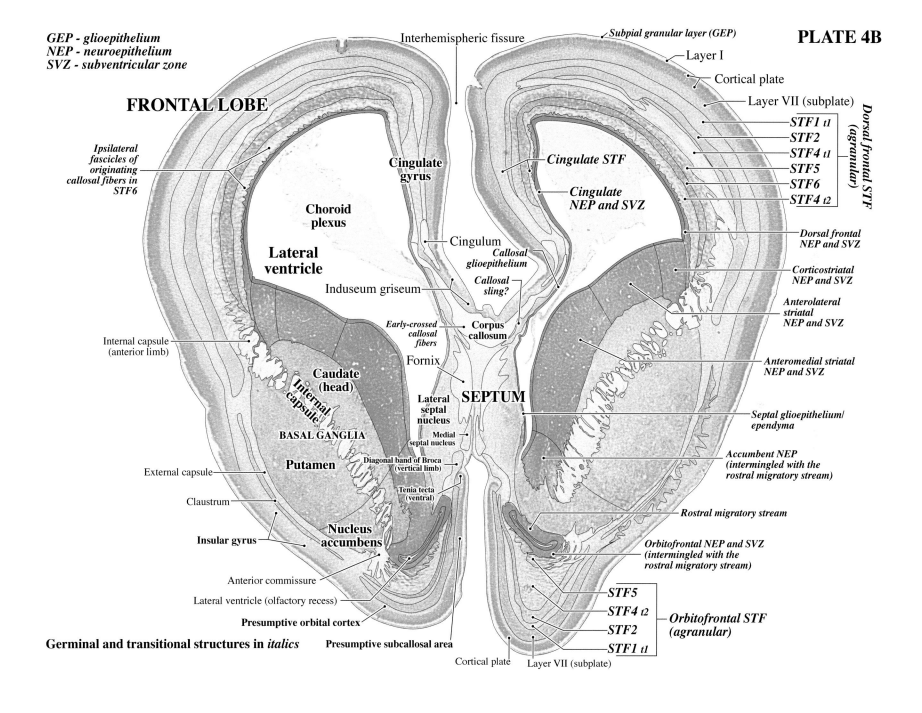

PLATE 4B

GEP - *glioepithelium*
NEP - *neuroepithelium*
SVZ - *subventricular zone*

Interhemispheric fissure

Subpial granular layer (GEP)

Layer I

Cortical plate

Layer VII (subplate)

FRONTAL LOBE

Dorsal frontal STF (agranular)

STF1 t1
STF2
STF4 t1
STF5
STF6
STF4 t2

Ipsilateral fascicles of originating callosal fibers in STF6

Cingulate STF

Cingulate gyrus

Cingulate NEP and SVZ

Choroid plexus

Cingulum

Dorsal frontal NEP and SVZ

Callosal glioepithelium

Lateral ventricle

Induseum griseum

Callosal sling?

Corticostriatal NEP and SVZ

Early-crossed callosal fibers

Corpus callosum

Anterolateral striatal NEP and SVZ

Internal capsule (anterior limb)

Fornix

Anteromedial striatal NEP and SVZ

Caudate (head)

Internal capsule

Lateral septal nucleus

SEPTUM

Septal glioepithelium/ ependyma

BASAL GANGLIA

Medial septal nucleus

External capsule

Putamen

Diagonal band of Broca (vertical limb)

Accumbent NEP (intermingled with the rostral migratory stream)

Claustrum

Tenia tecta (ventral)

Nucleus accumbens

Rostral migratory stream

Insular gyrus

Orbitofrontal NEP and SVZ (intermingled with the rostral migratory stream)

Anterior commissure

Lateral ventricle (olfactory recess)

STF5
STF4 t2
STF2
STF1 t1

Orbitofrontal STF (agranular)

Germinal and transitional structures in *italics*

Presumptive orbital cortex

Presumptive subcallosal area

Cortical plate

Layer VII (subplate)

18

PLATE 5A
CR 96 mm, GW 14, Y144-63
Frontal
Section 281

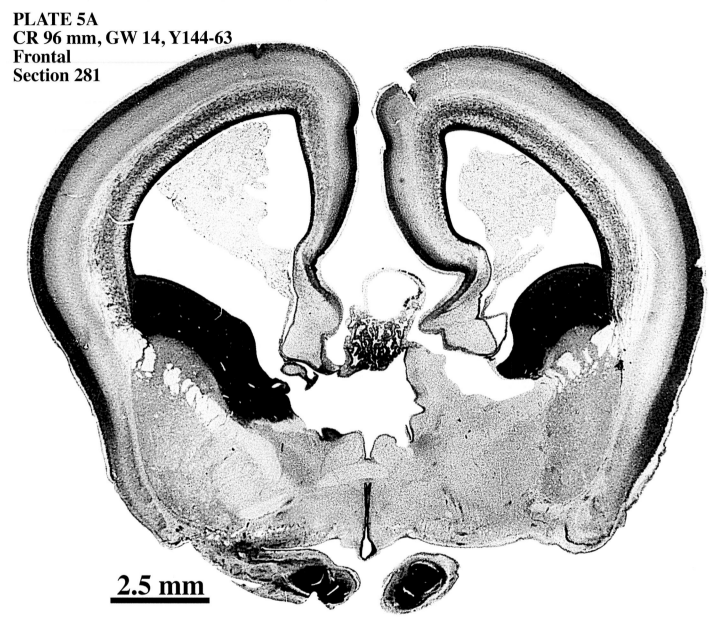

2.5 mm

LAYERS OF THE CORTICAL
STRATIFIED TRANSITIONAL
FIELD (STF)

STF1 Superficial fibrous layer with an early developmental stage *(t1)* when many cells are migrating through it, followed by a late stage *(t2)* with sparse cells. Endures as the subcortical white matter.

STF2 Upper cellular layer, the last sojourn zone before cells translocate to the cortical plate.

STF4 Complex middle layer with three developmental stages: *t1*– fibrous layer without interspersed cells; *t2*– cells and fibers intermingle to form striations; *t3*– fibers endure in the deep white matter.

STF5 Deep cellular layer, the first sojourn zone to appear outside the germinal matrix.

STF6 Late-forming deep layer of callosal fibers outside the germinal matrix.

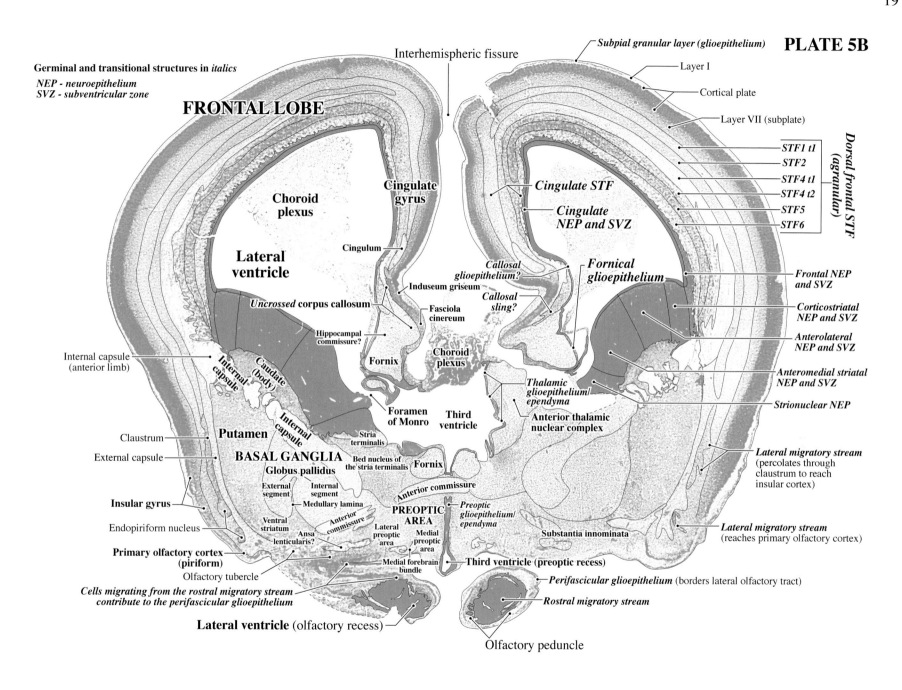

PLATE 5B

Germinal and transitional structures in *italics*

NEP - neuroepithelium
SVZ - subventricular zone

Subpial granular layer (glioepithelium)

Layer I

Cortical plate

Layer VII (subplate)

Interhemispheric fissure

FRONTAL LOBE

Dorsal frontal STF (agranular)

STF1 t1
STF2
STF4 t1
STF4 t2
STF5
STF6

Cingulate gyrus

Cingulate STF

Cingulate NEP and SVZ

Choroid plexus

Cingulum

Callosal glioepithelium?

Fornical glioepithelium

Lateral ventricle

Induseum griseum

Callosal sling?

Frontal NEP and SVZ

Uncrossed corpus callosum

Fasciola cinereum

Corticostriatal NEP and SVZ

Hippocampal commissure?

Fornix

Choroid plexus

Anterolateral NEP and SVZ

Internal capsule (anterior limb)

Internal capsule

Caudate (body)

Anteromedial striatal NEP and SVZ

Thalamic glioepithelium/ ependyma

Strionuclear NEP

Internal capsule

Foramen of Monro

Third ventricle

Anterior thalamic nuclear complex

Claustrum

Putamen

Stria terminalis

BASAL GANGLIA

Bed nucleus of the stria terminalis

Fornix

Lateral migratory stream (percolates through claustrum to reach insular cortex)

External capsule

Globus pallidus

Anterior commissure

External segment

Internal segment

Medullary lamina

Insular gyrus

Ventral striatum

Anterior commissure

PREOPTIC AREA

Preoptic glioepithelium/ ependyma

Substantia innominata

Lateral migratory stream (reaches primary olfactory cortex)

Endopiriform nucleus

Ansa lenticularis?

Lateral preoptic area

Medial preoptic area

Primary olfactory cortex (piriform)

Medial forebrain bundle

Third ventricle (preoptic recess)

Olfactory tubercle

Perifascicular glioepithelium (borders lateral olfactory tract)

Cells migrating from the rostral migratory stream contribute to the perifascicular glioepithelium

Rostral migratory stream

Lateral ventricle (olfactory recess)

Olfactory peduncle

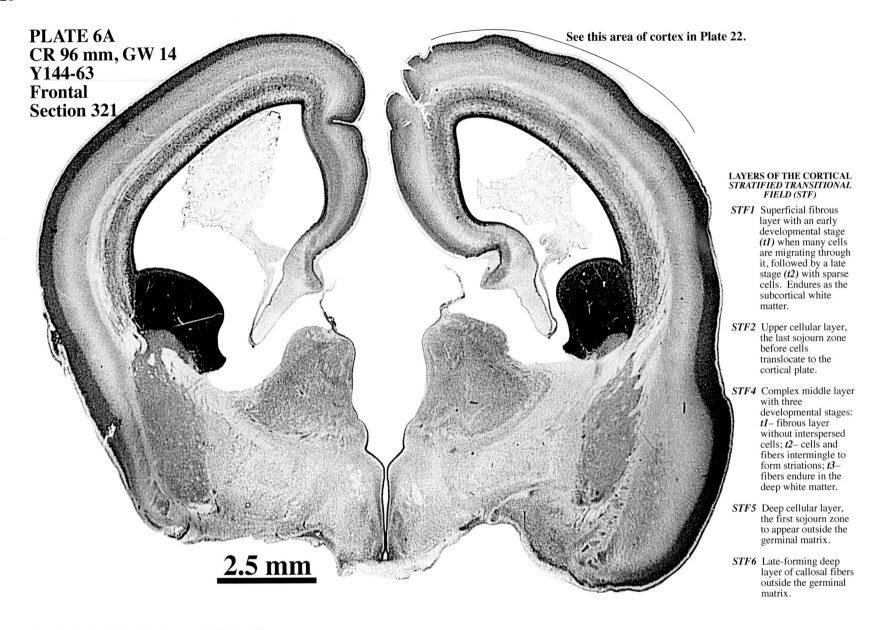

PLATE 6A
CR 96 mm, GW 14
Y144-63
Frontal
Section 321

See this area of cortex in Plate 22.

2.5 mm

LAYERS OF THE CORTICAL
STRATIFIED TRANSITIONAL
FIELD (STF)

STF1 Superficial fibrous layer with an early developmental stage *(t1)* when many cells are migrating through it, followed by a late stage *(t2)* with sparse cells. Endures as the subcortical white matter.

STF2 Upper cellular layer, the last sojourn zone before cells translocate to the cortical plate.

STF4 Complex middle layer with three developmental stages: *t1*– fibrous layer without interspersed cells; *t2*– cells and fibers intermingle to form striations; *t3*– fibers endure in the deep white matter.

STF5 Deep cellular layer, the first sojourn zone to appear outside the germinal matrix.

STF6 Late-forming deep layer of callosal fibers outside the germinal matrix.

See detail of the thalamus in Plate 17.

PLATE 6B

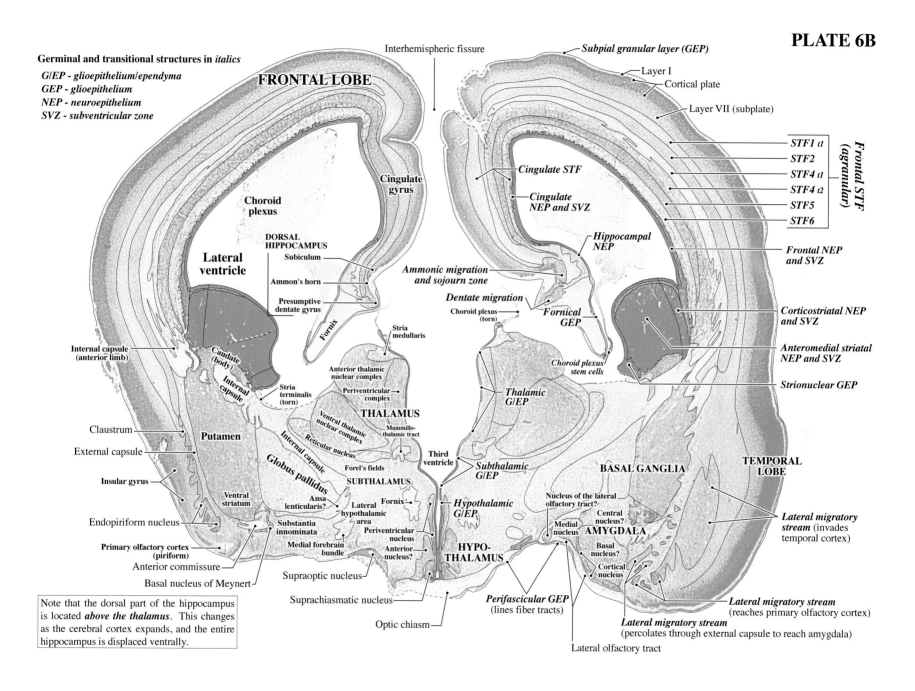

Germinal and transitional structures in *italics*

G/EP - glioepithelium/ependyma
GEP - glioepithelium
NEP - neuroepithelium
SVZ - subventricular zone

Interhemispheric fissure

Subpial granular layer (GEP)

FRONTAL LOBE

Layer I
Cortical plate

Layer VII (subplate)

STF1 t1
STF2
STF4 t1
STF4 t2
STF5
STF6

Frontal STF (agranular)

Cingulate gyrus

Cingulate STF

Cingulate NEP and SVZ

Choroid plexus

Hippocampal NEP

DORSAL HIPPOCAMPUS

Lateral ventricle

Subiculum

Ammonic migration and sojourn zone

Frontal NEP and SVZ

Ammon's horn

Presumptive dentate gyrus

Dentate migration

Choroid plexus (torn)

Fornical GEP

Corticostriatal NEP and SVZ

Fornix

Stria medullaris

Choroid plexus stem cells

Anteromedial striatal NEP and SVZ

Internal capsule (anterior limb)

Caudate (body)

Internal capsule

Stria terminalis (torn)

Anterior thalamic nuclear complex

Periventricular complex

Thalamic G/EP

Strionuclear GEP

Claustrum

Putamen

THALAMUS

Ventral thalamic nuclear complex

Mammillo-thalamic tract

External capsule

Internal capsule

Reticular nucleus

Third ventricle

Subthalamic G/EP

BASAL GANGLIA

TEMPORAL LOBE

Insular gyrus

Globus pallidus

Forel's fields

SUBTHALAMUS

Ventral striatum

Ansa lenticularis?

Lateral hypothalamic area

Fornix

Hypothalamic G/EP

Nucleus of the lateral olfactory tract?

Central nucleus?

Lateral migratory stream (invades temporal cortex)

Endopiriform nucleus

Substantia innominata

Periventricular nucleus

Medial nucleus

AMYGDALA

Primary olfactory cortex (piriform)

Medial forebrain bundle

Anterior nucleus?

Basal nucleus?

Anterior commissure

HYPO-THALAMUS

Cortical nucleus

Lateral migratory stream (reaches primary olfactory cortex)

Basal nucleus of Meynert

Supraoptic nucleus

Suprachiasmatic nucleus

Perifascicular GEP (lines fiber tracts)

Lateral migratory stream (percolates through external capsule to reach amygdala)

Optic chiasm

Lateral olfactory tract

Note that the dorsal part of the hippocampus is located **above the thalamus**. This changes as the cerebral cortex expands, and the entire hippocampus is displaced ventrally.

PLATE 7A
CR 96 mm, GW 14
Y144-63
Frontal
Section 361

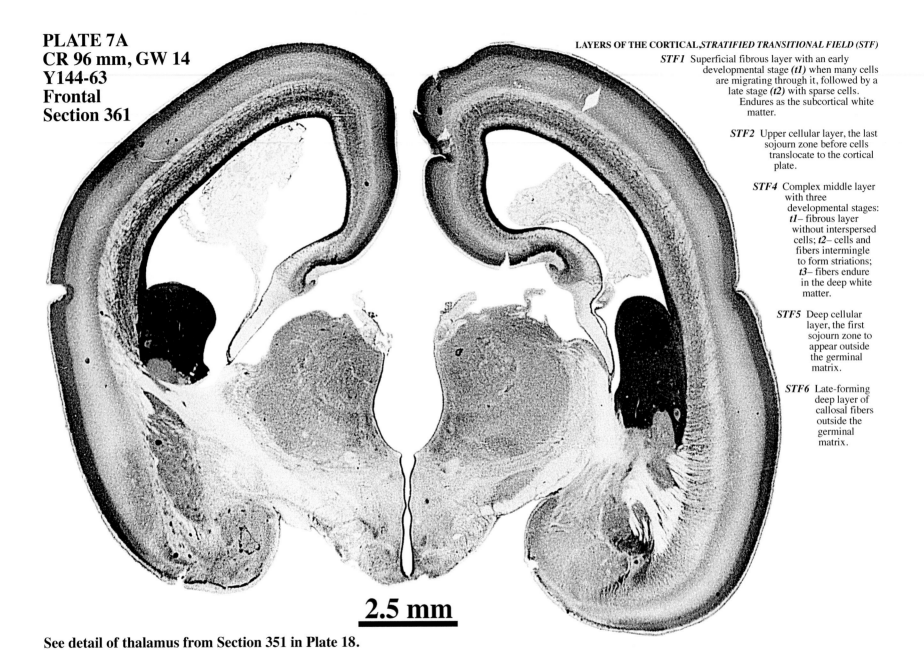

LAYERS OF THE CORTICAL *STRATIFIED TRANSITIONAL FIELD (STF)*

STF1 Superficial fibrous layer with an early developmental stage *(t1)* when many cells are migrating through it, followed by a late stage *(t2)* with sparse cells. Endures as the subcortical white matter.

STF2 Upper cellular layer, the last sojourn zone before cells translocate to the cortical plate.

STF4 Complex middle layer with three developmental stages: *t1*– fibrous layer without interspersed cells; *t2*– cells and fibers intermingle to form striations; *t3*– fibers endure in the deep white matter.

STF5 Deep cellular layer, the first sojourn zone to appear outside the germinal matrix.

STF6 Late-forming deep layer of callosal fibers outside the germinal matrix.

2.5 mm

See detail of thalamus from Section 351 in Plate 18.

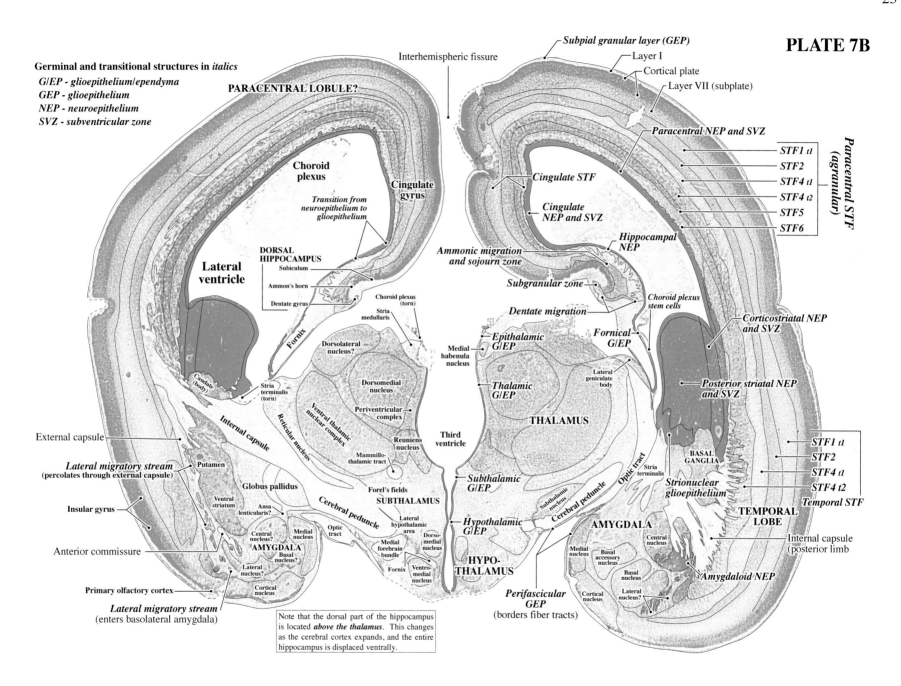

23

PLATE 7B

Germinal and transitional structures in *italics*

G/EP - glioepithelium/ependyma
GEP - glioepithelium
NEP - neuroepithelium
SVZ - subventricular zone

Interhemispheric fissure

Subpial granular layer (GEP)
Layer I
Cortical plate
Layer VII (subplate)

PARACENTRAL LOBULE?

Paracentral NEP and SVZ

Choroid plexus

Cingulate gyrus

Cingulate STF

STF1 t1
STF2
STF4 t1
STF4 t2
STF5
STF6

Paracentral STF (agranular)

Transition from neuroepithelium to glioepithelium

Cingulate NEP and SVZ

DORSAL HIPPOCAMPUS

Hippocampal NEP

Subiculum

Ammonic migration and sojourn zone

Ammon's horn

Lateral ventricle

Dentate gyrus

Choroid plexus (torn)

Subgranular zone

Choroid plexus stem cells

Stria medullaris

Dentate migration

Corticostriatal NEP and SVZ

Fornix

Dorsolateral nucleus?

Epithalamic G/EP

Fornical G/EP

Medial habenula nucleus

Caudate (body)

Stria terminalis (torn)

Dorsomedial nucleus

Thalamic G/EP

Lateral geniculate body

Posterior striatal NEP and SVZ

Periventricular complex

THALAMUS

External capsule

Internal capsule

Ventral thalamic nuclear complex

Reuniens nucleus

Third ventricle

BASAL GANGLIA

Reticular nucleus

STF1 t1
STF2
STF4 t1
STF4 t2

Lateral migratory stream (percolates through external capsule)

Putamen

Mammillo-thalamic tract

Subthalamic G/EP

Stria terminalis

Strionuclear glioepithelium

Temporal STF

Globus pallidus

Forel's fields

Optic tract

Insular gyrus

Ventral striatum

SUBTHALAMUS

Subthalamic nucleus

Cerebral peduncle

TEMPORAL LOBE

Ansa lenticularis?

Cerebral peduncle

Lateral hypothalamic area

Hypothalamic G/EP

AMYGDALA

Internal capsule (posterior limb

Optic tract

Medial nucleus

Anterior commissure

Central nucleus?

Medial nucleus

Dorso-medial nucleus

Central nucleus

AMYGDALA

Basal nucleus?

Medial forebrain bundle

Basal accessory nucleus

Lateral nucleus?

HYPO-THALAMUS

Basal nucleus

Amygdaloid NEP

Primary olfactory cortex

Cortical nucleus

Fornix

Ventro-medial nucleus

Perifascicular GEP (borders fiber tracts)

Cortical nucleus

Lateral nucleus?

Lateral migratory stream (enters basolateral amygdala)

Note that the dorsal part of the hippocampus is located *above the thalamus*. This changes as the cerebral cortex expands, and the entire hippocampus is displaced ventrally.

PLATE 8A
CR 96 mm, GW 14
Y144-63
Frontal
Section 401

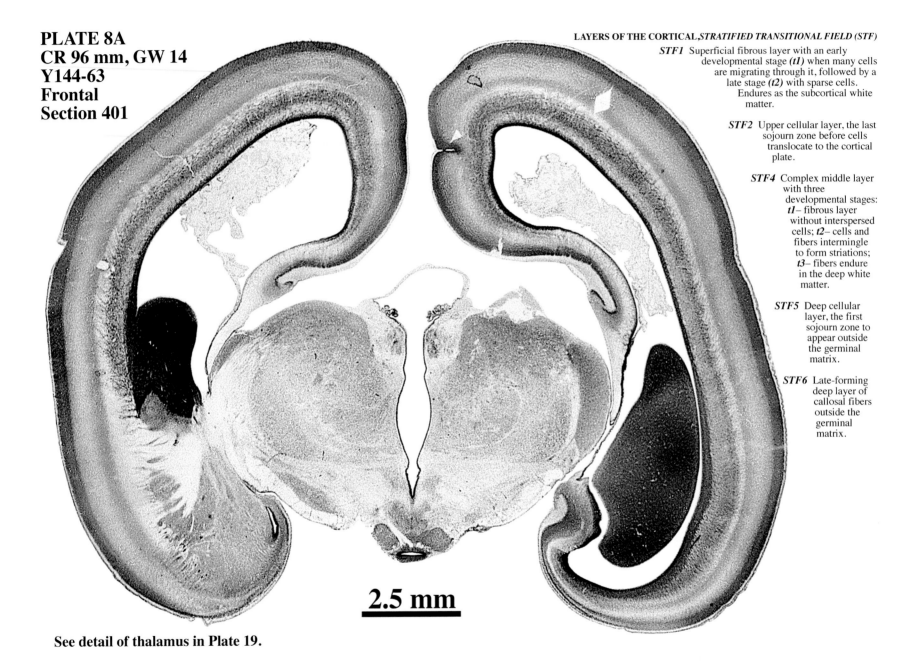

2.5 mm

See detail of thalamus in Plate 19.

LAYERS OF THE CORTICAL *STRATIFIED TRANSITIONAL FIELD (STF)*

STF1 Superficial fibrous layer with an early developmental stage *(t1)* when many cells are migrating through it, followed by a late stage *(t2)* with sparse cells. Endures as the subcortical white matter.

STF2 Upper cellular layer, the last sojourn zone before cells translocate to the cortical plate.

STF4 Complex middle layer with three developmental stages: *t1*– fibrous layer without interspersed cells; *t2*– cells and fibers intermingle to form striations; *t3*– fibers endure in the deep white matter.

STF5 Deep cellular layer, the first sojourn zone to appear outside the germinal matrix.

STF6 Late-forming deep layer of callosal fibers outside the germinal matrix.

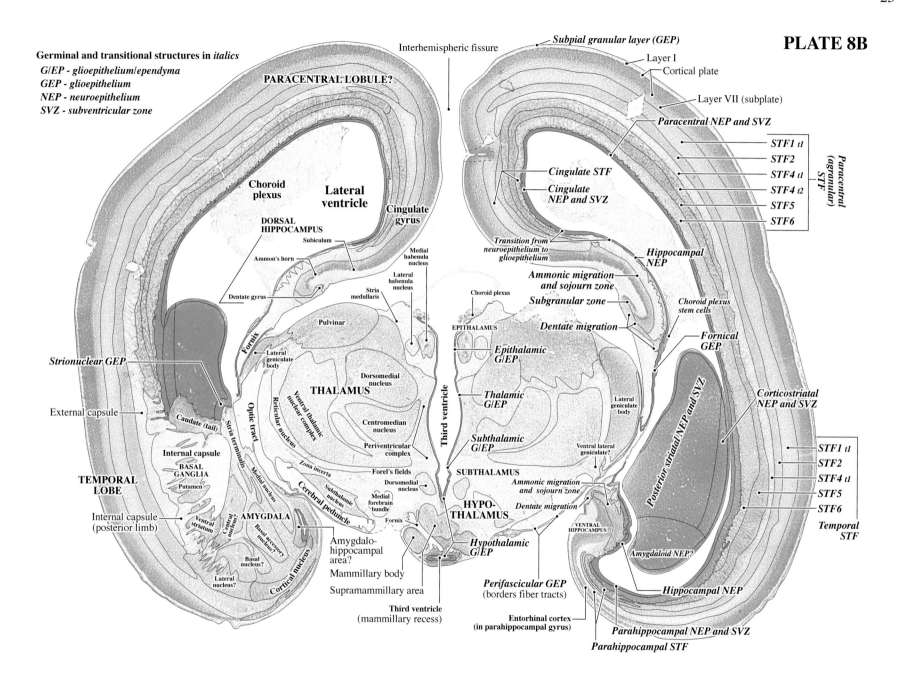

PLATE 8B

Germinal and transitional structures in *italics*

G/EP - glioepithelium/ependyma
GEP - glioepithelium
NEP - neuroepithelium
SVZ - subventricular zone

Interhemispheric fissure

Subpial granular layer (GEP)
Layer I
Cortical plate
Layer VII (subplate)
Paracentral NEP and SVZ

PARACENTRAL LOBULE?

STF1 t1
STF2
STF4 t1
STF4 t2
STF5
STF6

Paracentral (agranular) STF

Choroid plexus

Lateral ventricle

Cingulate gyrus

Cingulate STF
Cingulate NEP and SVZ

DORSAL HIPPOCAMPUS

Subiculum

Medial habenula nucleus

Transition from neuroepithelium to glioepithelium

Hippocampal NEP

Ammon's horn

Lateral habenula nucleus

Ammonic migration and sojourn zone

Dentate gyrus

Stria medullaris

Subgranular zone

Choroid plexus stem cells

Choroid plexus

Dentate migration

Fornical GEP

Strionuclear GEP

Fornix

Lateral geniculate body

Pulvinar

EPITHALAMUS

Epithalamic G/EP

External capsule

Dorsomedial nucleus

THALAMUS

Thalamic G/EP

Lateral geniculate body

Corticostriatal NEP and SVZ

Caudate (tail)

Optic tract

Ventral thalamic nuclear complex

Reticular nucleus

Centromedian nucleus

Stria terminalis

Subthalamic G/EP

Periventricular complex

Internal capsule

BASAL GANGLIA

Medial nucleus

Zona incerta

Forel's fields

SUBTHALAMUS

STF1 t1
STF2
STF4 t1
STF5
STF6

Ventral lateral geniculate?

Posterior striatal NEP and SVZ

TEMPORAL LOBE

Putamen

Cerebral peduncle

Subthalamic nucleus

Dorsomedial nucleus

HYPO-THALAMUS

Ammonic migration and sojourn zone

Dentate migration

Temporal STF

Internal capsule (posterior limb)

Ventral striatum

Central nucleus?

AMYGDALA

Basal accessory nucleus?

Medial forebrain bundle

Fornix

VENTRAL HIPPOCAMPUS

Amygdaloid NEP?

Basal nucleus?

Lateral nucleus?

Cortical nucleus

Amygdalo-hippocampal area?

Mammillary body

Supramammillary area

Hypothalamic G/EP

Perifascicular GEP (borders fiber tracts)

Hippocampal NEP

Third ventricle (mammillary recess)

Entorhinal cortex (in parahippocampal gyrus)

Parahippocampal NEP and SVZ

Parahippocampal STF

Third ventricle

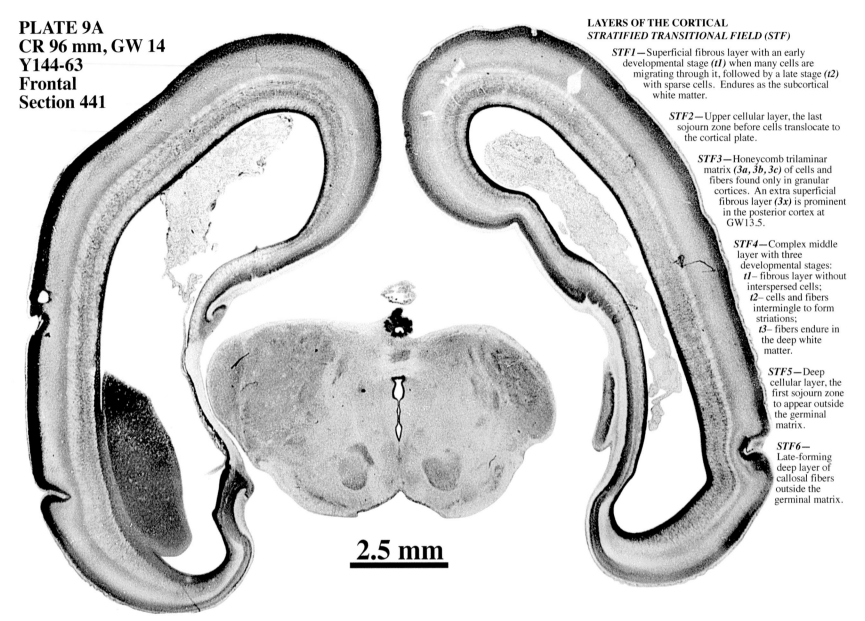

PLATE 9A
CR 96 mm, GW 14
Y144-63
Frontal
Section 441

LAYERS OF THE CORTICAL
STRATIFIED TRANSITIONAL FIELD (STF)

STF1—Superficial fibrous layer with an early developmental stage *(t1)* when many cells are migrating through it, followed by a late stage *(t2)* with sparse cells. Endures as the subcortical white matter.

STF2—Upper cellular layer, the last sojourn zone before cells translocate to the cortical plate.

STF3—Honeycomb trilaminar matrix *(3a, 3b, 3c)* of cells and fibers found only in granular cortices. An extra superficial fibrous layer *(3x)* is prominent in the posterior cortex at GW13.5.

STF4—Complex middle layer with three developmental stages:
   t1– fibrous layer without interspersed cells;
   t2– cells and fibers intermingle to form striations;
   t3– fibers endure in the deep white matter.

STF5—Deep cellular layer, the first sojourn zone to appear outside the germinal matrix.

STF6—Late-forming deep layer of callosal fibers outside the germinal matrix.

2.5 mm

See detail of thalamus from Section 421 in Plate 20.

PLATE 9B

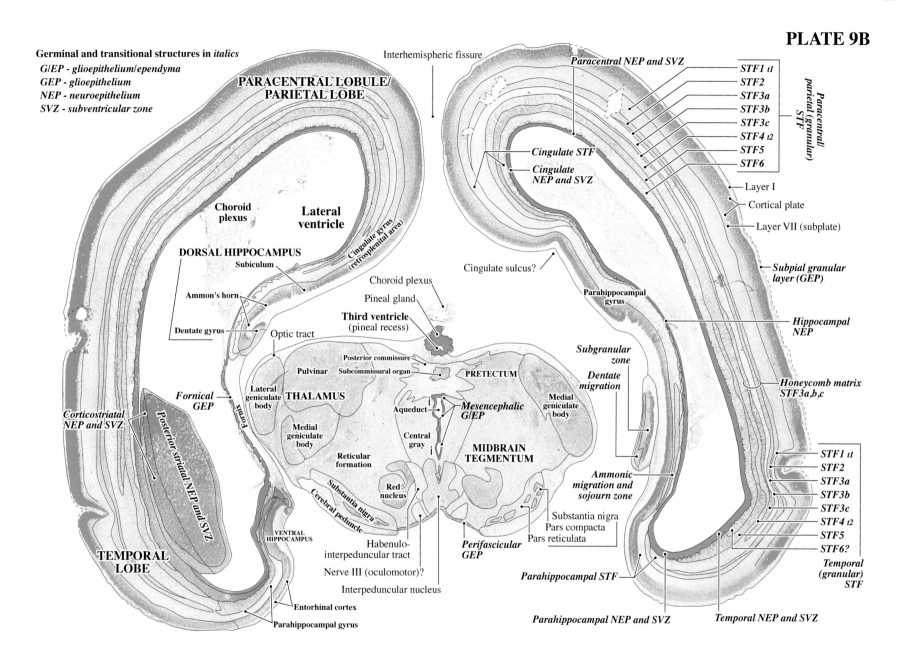

Germinal and transitional structures in *italics*
G/EP - glioepithelium/ependyma
GEP - glioepithelium
NEP - neuroepithelium
SVZ - subventricular zone

Interhemispheric fissure

Paracentral NEP and SVZ

STF1 *t1*
STF2
STF3a
STF3b
STF3c
STF4 *t2*
STF5
STF6

Paracentral/ parietal (granular) STF

PARACENTRAL LOBULE/ PARIETAL LOBE

Cingulate STF
Cingulate NEP and SVZ

Layer I
Cortical plate
Layer VII (subplate)

Choroid plexus

Lateral ventricle

Cingulate gyrus (retrosplenital area)

Cingulate sulcus?

Subpial granular layer (GEP)

DORSAL HIPPOCAMPUS
Subiculum

Choroid plexus

Pineal gland

Parahippocampal gyrus

Ammon's horn

Hippocampal NEP

Dentate gyrus

Optic tract

Third ventricle (pineal recess)

Subgranular zone

Honeycomb matrix STF3a,b,c

Posterior commissure
Subcommissural organ

Pulvinar

Dentate migration

Fornical GEP

Lateral geniculate body

THALAMUS

PRETECTUM

Aqueduct

Mesencephalic G/EP

Medial geniculate body

Corticostriatal NEP and SVZ

Posterior striatal NEP and SVZ

Medial geniculate body

Central gray

MIDBRAIN TEGMENTUM

Reticular formation

Ammonic migration and sojourn zone

STF1 *t1*
STF2
STF3a
STF3b
STF3c
STF4 *t2*
STF5
STF6?

Substantia nigra

Red nucleus

Substantia nigra
Pars compacta
Pars reticulata

Temporal (granular) STF

VENTRAL HIPPOCAMPUS

Cerebral peduncle

Habenulo-interpeduncular tract

Perifascicular GEP

TEMPORAL LOBE

Nerve III (oculomotor)?

Interpeduncular nucleus

Parahippocampal STF

Entorhinal cortex

Parahippocampal gyrus

Parahippocampal NEP and SVZ

Temporal NEP and SVZ

**PLATE 10A
CR 96 mm, GW 14
Y144-63
Frontal
Section 481**

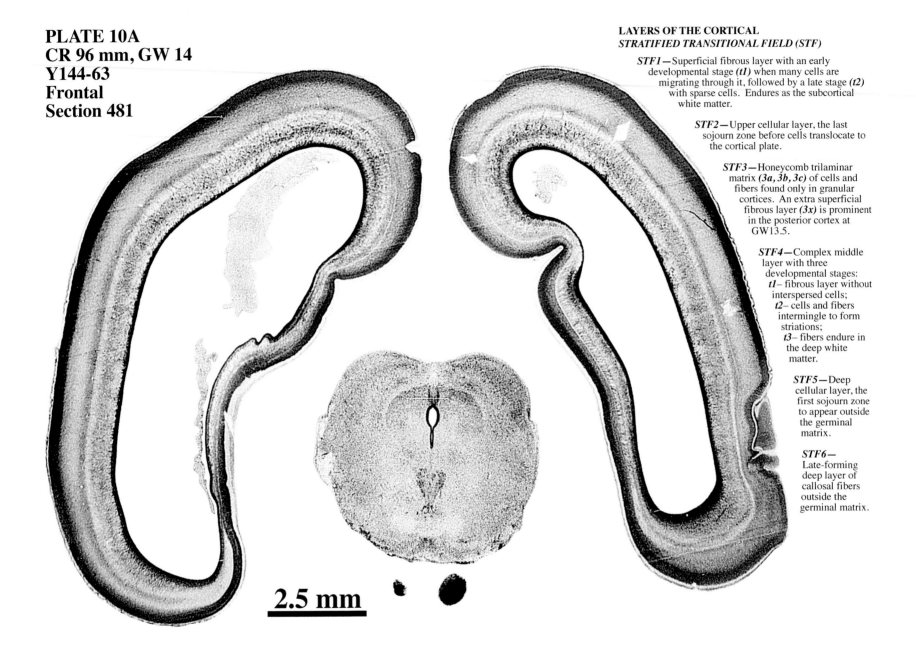

2.5 mm

**LAYERS OF THE CORTICAL
*STRATIFIED TRANSITIONAL FIELD (STF)***

STF1—Superficial fibrous layer with an early developmental stage *(t1)* when many cells are migrating through it, followed by a late stage *(t2)* with sparse cells. Endures as the subcortical white matter.

STF2—Upper cellular layer, the last sojourn zone before cells translocate to the cortical plate.

STF3—Honeycomb trilaminar matrix *(3a, 3b, 3c)* of cells and fibers found only in granular cortices. An extra superficial fibrous layer *(3x)* is prominent in the posterior cortex at GW13.5.

STF4—Complex middle layer with three developmental stages:
t1– fibrous layer without interspersed cells;
t2– cells and fibers intermingle to form striations;
t3– fibers endure in the deep white matter.

STF5—Deep cellular layer, the first sojourn zone to appear outside the germinal matrix.

STF6—Late-forming deep layer of callosal fibers outside the germinal matrix.

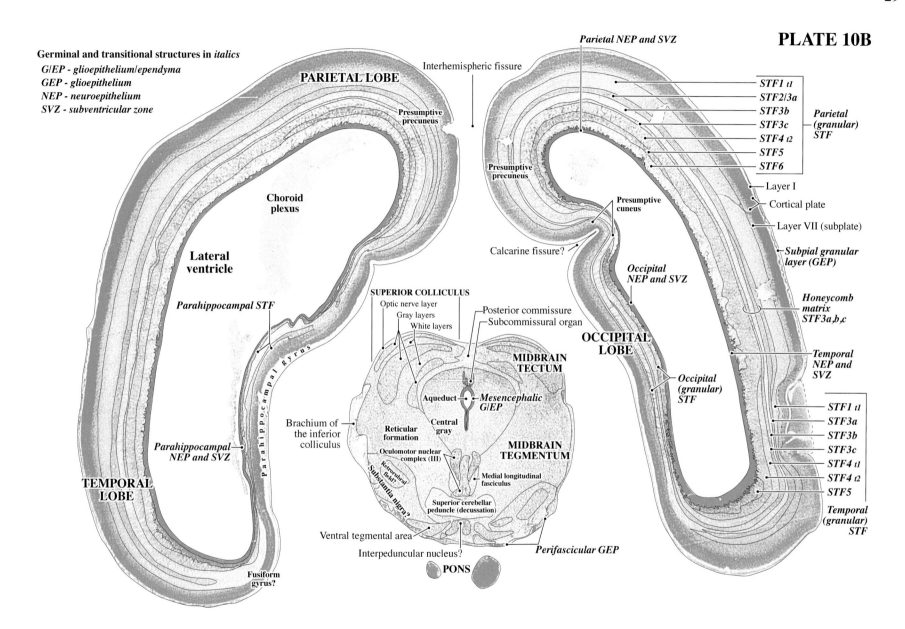

PLATE 10B

Germinal and transitional structures in *italics*

G/EP - glioepithelium/ependyma
GEP - glioepithelium
NEP - neuroepithelium
SVZ - subventricular zone

Parietal NEP and SVZ

PARIETAL LOBE

Interhemispheric fissure

Presumptive
precuneus

STF1 *t1*
STF2/3a
STF3b
STF3c
STF4 *t2*
STF5
STF6

*Parietal
(granular)
STF*

Presumptive
precuneus

Layer I
Cortical plate
Layer VII (subplate)

**Choroid
plexus**

Presumptive
cuneus

*Subpial granular
layer (GEP)*

Calcarine fissure?

*Occipital
NEP and SVZ*

**Lateral
ventricle**

*Honeycomb
matrix
STF3a,b,c*

Parahippocampal STF

SUPERIOR COLLICULUS

Optic nerve layer
Gray layers
White layers

Posterior commissure
Subcommissural organ

**MIDBRAIN
TECTUM**

**OCCIPITAL
LOBE**

*Occipital
(granular)
STF*

*Temporal
NEP and
SVZ*

P a r a h i p p o c a m p a l   g y r u s

Aqueduct

*Mesencephalic
G/EP*

Brachium of
the inferior
colliculus

**Reticular
formation**

**Central
gray**

**MIDBRAIN
TEGMENTUM**

STF1 *t1*
STF3a
STF3b
STF3c
STF4 *t1*
STF4 *t2*
STF5

Oculomotor nuclear
complex (III)

*Retrorubral
field?*

Substantia nigra?

Medial longitudinal
fasciculus

*Parahippocampal
NEP and SVZ*

**TEMPORAL
LOBE**

Superior cerebellar
peduncle (decussation)

*Temporal
(granular)
STF*

Ventral tegmental area

Interpeduncular nucleus?

Perifascicular GEP

Fusiform
gyrus?

PONS

PLATE 11A
CR 96 mm, GW 14
Y144-63
Frontal
Section 521

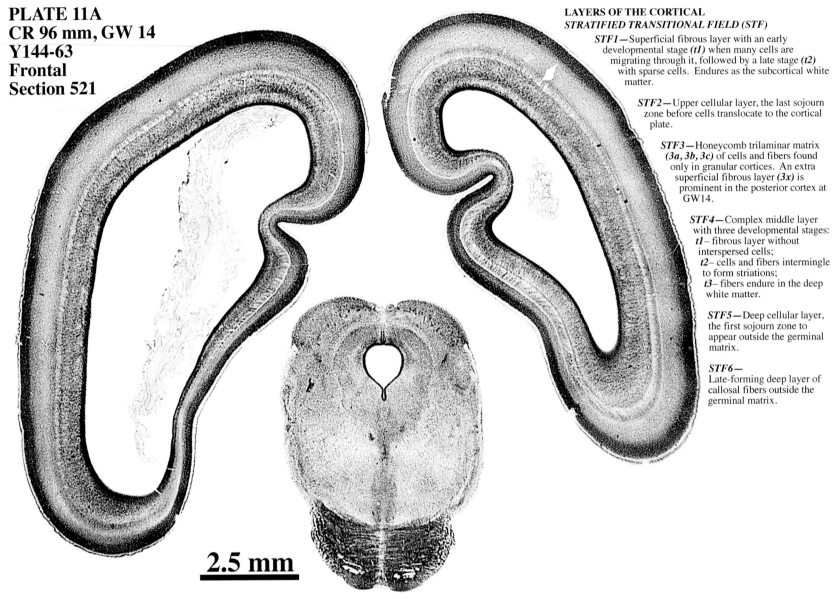

2.5 mm

See details of the cortex
in Plates 23 and 24.

LAYERS OF THE CORTICAL
STRATIFIED TRANSITIONAL FIELD (STF)

STF1—Superficial fibrous layer with an early developmental stage *(t1)* when many cells are migrating through it, followed by a late stage *(t2)* with sparse cells. Endures as the subcortical white matter.

STF2—Upper cellular layer, the last sojourn zone before cells translocate to the cortical plate.

STF3—Honeycomb trilaminar matrix *(3a, 3b, 3c)* of cells and fibers found only in granular cortices. An extra superficial fibrous layer *(3x)* is prominent in the posterior cortex at GW14.

STF4—Complex middle layer with three developmental stages:
t1– fibrous layer without interspersed cells;
t2– cells and fibers intermingle to form striations;
t3– fibers endure in the deep white matter.

STF5—Deep cellular layer, the first sojourn zone to appear outside the germinal matrix.

STF6—
Late-forming deep layer of callosal fibers outside the germinal matrix.

PLATE 11B

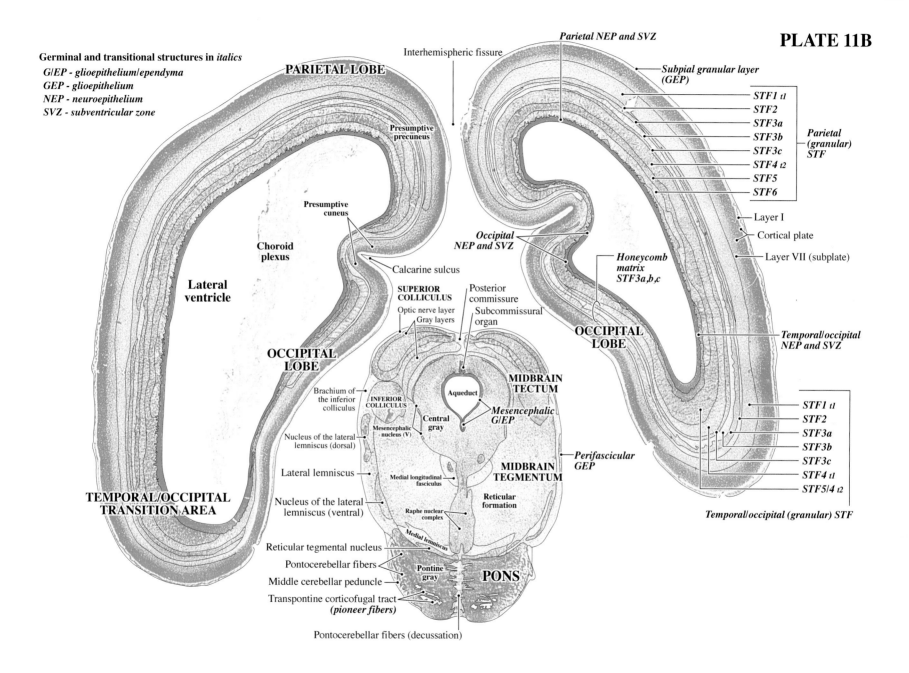

Germinal and transitional structures in *italics*
G/EP - glioepithelium/ependyma
GEP - glioepithelium
NEP - neuroepithelium
SVZ - subventricular zone

PARIETAL LOBE

Interhemispheric fissure

Parietal NEP and SVZ

Subpial granular layer (GEP)

STF1 t1
STF2
STF3a
STF3b
STF3c
STF4 t2
STF5
STF6

Parietal (granular) STF

Presumptive precuneus

- Layer I
- Cortical plate

Presumptive cuneus

Occipital NEP and SVZ

- Layer VII (subplate)

Choroid plexus

Calcarine sulcus

Honeycomb matrix STF3a,b,c

OCCIPITAL LOBE

Temporal/occipital NEP and SVZ

Lateral ventricle

SUPERIOR COLLICULUS
Optic nerve layer
Gray layers

Posterior commissure
Subcommissural organ

OCCIPITAL LOBE

MIDBRAIN TECTUM

Aqueduct

Mesencephalic G/EP

Brachium of the inferior colliculus

INFERIOR COLLICULUS

Central gray

Mesencephalic nucleus (V)

Nucleus of the lateral lemniscus (dorsal)

Lateral lemniscus

Medial longitudinal fasciculus

MIDBRAIN TEGMENTUM

Perifascicular GEP

Reticular formation

Nucleus of the lateral lemniscus (ventral)

STF1 t1
STF2
STF3a
STF3b
STF3c
STF4 t1
STF5/4 t2

Temporal/occipital (granular) STF

Raphe nuclear complex

Reticular tegmental nucleus

Medial lemniscus

Pontocerebellar fibers

Middle cerebellar peduncle

Pontine gray

PONS

Transpontine corticofugal tract
(pioneer fibers)

TEMPORAL/OCCIPITAL TRANSITION AREA

Pontocerebellar fibers (decussation)

PLATE 12A
CR 96 mm, GW 14
Y144-63
Frontal
Section 561

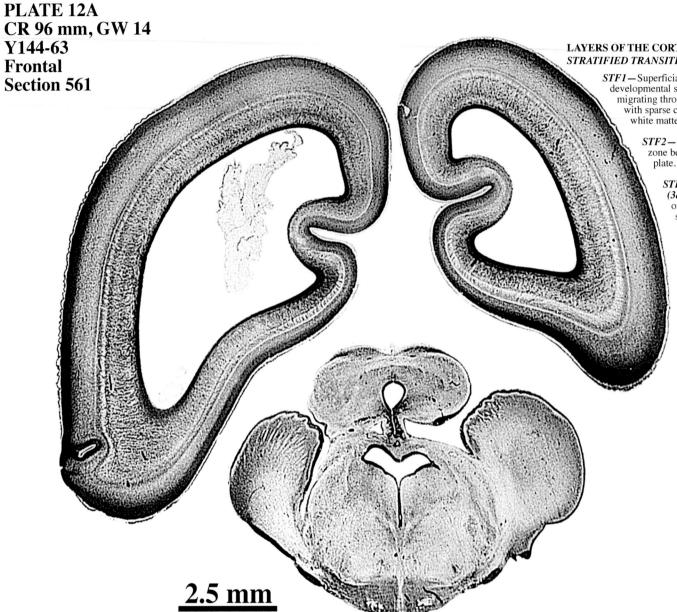

2.5 mm

LAYERS OF THE CORTICAL
STRATIFIED TRANSITIONAL FIELD (STF)

STF1—Superficial fibrous layer with an early developmental stage *(t1)* when many cells are migrating through it, followed by a late stage *(t2)* with sparse cells. Endures as the subcortical white matter.

STF2—Upper cellular layer, the last sojourn zone before cells translocate to the cortical plate.

STF3—Honeycomb trilaminar matrix *(3a, 3b, 3c)* of cells and fibers found only in granular cortices. An extra superficial fibrous layer *(3x)* is prominent in the posterior cortex at GW14.

STF4—Complex middle layer with three developmental stages:
t1– fibrous layer without interspersed cells;
t2– cells and fibers intermingle to form striations;
t3– fibers endure in the deep white matter.

STF5—Deep cellular layer, the first sojourn zone to appear outside the germinal matrix.

STF6—
Late-forming deep layer of callosal fibers outside the germinal matrix.

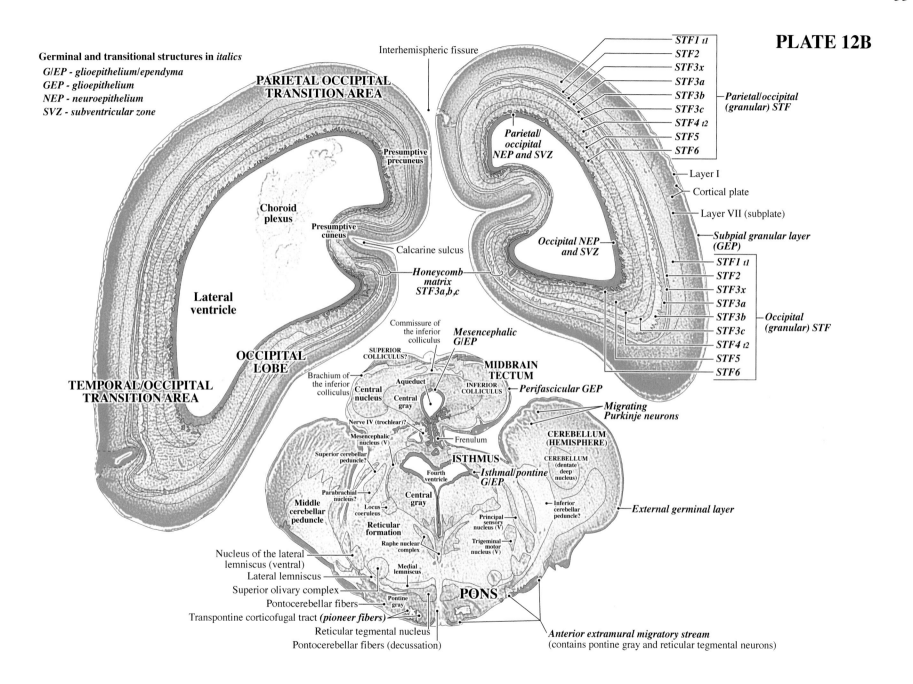

PLATE 12B

Germinal and transitional structures in *italics*

G/EP - glioepithelium/ependyma
GEP - glioepithelium
NEP - neuroepithelium
SVZ - subventricular zone

Interhemispheric fissure

PARIETAL OCCIPITAL TRANSITION AREA

Presumptive precuneus

STF1 t1
STF2
STF3x
STF3a
STF3b
STF3c
STF4 t2
STF5
STF6

Parietal/occipital (granular) STF

Parietal/occipital NEP and SVZ

Layer I
Cortical plate
Layer VII (subplate)

Choroid plexus

Presumptive cuneus

Calcarine sulcus

Occipital NEP and SVZ

Subpial granular layer (GEP)

Lateral ventricle

Honeycomb matrix STF3a,b,c

STF1 t1
STF2
STF3x
STF3a
STF3b
STF3c
STF4 t2
STF5
STF6

Occipital (granular) STF

OCCIPITAL LOBE

Commissure of the inferior colliculus

Mesencephalic G/EP

MIDBRAIN TECTUM

SUPERIOR COLLICULUS?

Brachium of the inferior colliculus

Central nucleus

Aqueduct

Central gray

INFERIOR COLLICULUS

Perifascicular GEP

TEMPORAL/OCCIPITAL TRANSITION AREA

Migrating Purkinje neurons

Nerve IV (trochlear)?

Mesencephalic nucleus (V)

Frenulum

CEREBELLUM (HEMISPHERE)

Superior cerebellar peduncle?

ISTHMUS

Isthmal/pontine G/EP

CEREBELLUM (dentate deep nucleus)

Parabrachial nucleus?

Fourth ventricle

Central gray

Locus coeruleus

Inferior cerebellar peduncle?

Middle cerebellar peduncle

Reticular formation

Principal sensory nucleus (V)

External germinal layer

Raphe nuclear complex

Trigeminal motor nucleus (V)

Nucleus of the lateral lemniscus (ventral)

Medial lemniscus

Lateral lemniscus

Superior olivary complex

Pontocerebellar fibers

Pontine gray

PONS

Transpontine corticofugal tract *(pioneer fibers)*

Reticular tegmental nucleus

Pontocerebellar fibers (decussation)

Anterior extramural migratory stream (contains pontine gray and reticular tegmental neurons)

PLATE 13A
CR 96 mm, GW 14
Y144-63
Frontal
Section 601

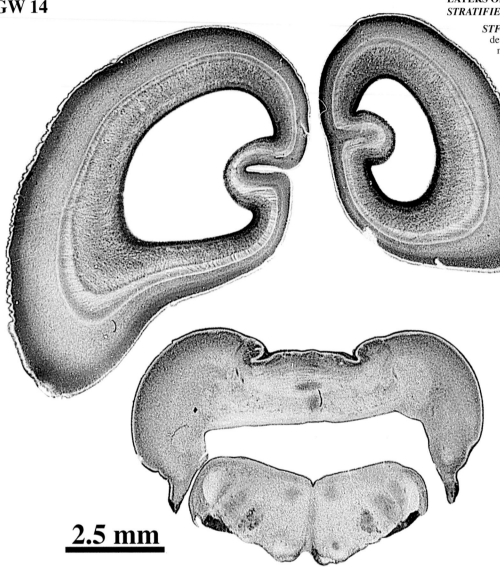

2.5 mm

LAYERS OF THE CORTICAL
STRATIFIED TRANSITIONAL FIELD (STF)

STF1—Superficial fibrous layer with an early developmental stage *(t1)* when many cells are migrating through it, followed by a late stage *(t2)* with sparse cells. Endures as the subcortical white matter.

STF2—Upper cellular layer, the last sojourn zone before cells translocate to the cortical plate.

STF3—Honeycomb trilaminar matrix *(3a, 3b, 3c)* of cells and fibers found only in granular cortices. An extra superficial fibrous layer *(3x)* is prominent in the posterior cortex at GW14.

STF4—Complex middle layer with three developmental stages:
t1– fibrous layer without interspersed cells;
t2– cells and fibers intermingle to form striations;
t3– fibers endure in the deep white matter.

STF5—Deep cellular layer, the first sojourn zone to appear outside the germinal matrix.

STF6—
Late-forming deep layer of callosal fibers outside the germinal matrix.

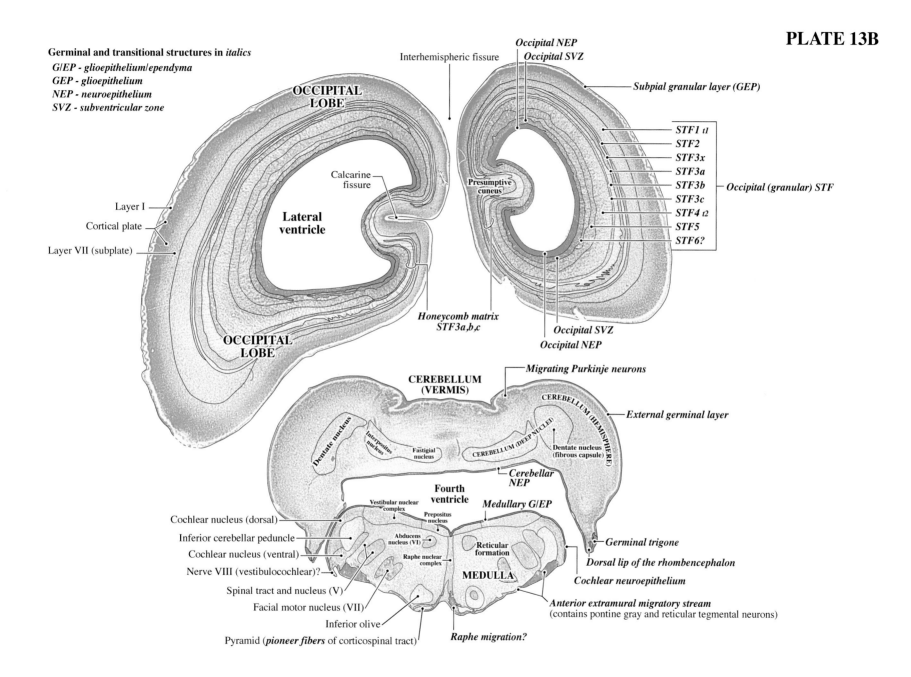

Germinal and transitional structures in *italics*
  G/EP - glioepithelium/ependyma
  GEP - glioepithelium
  NEP - neuroepithelium
  SVZ - subventricular zone

Occipital NEP
Occipital SVZ

Interhemispheric fissure

Subpial granular layer (GEP)

OCCIPITAL
LOBE

STF1 t1
STF2
STF3x
STF3a
STF3b
STF3c
STF4 t2
STF5
STF6?

Occipital (granular) STF

Calcarine
fissure

Presumptive
cuneus

Layer I

Cortical plate

**Lateral
ventricle**

Layer VII (subplate)

*Honeycomb matrix
STF3a,b,c*

Occipital SVZ
Occipital NEP

**OCCIPITAL
LOBE**

Migrating Purkinje neurons

**CEREBELLUM
(VERMIS)**

CEREBELLUM (HEMISPHERE)

External germinal layer

Dentate nucleus

Interpositus
nucleus

Fastigial
nucleus

CEREBELLUM (DEEP NUCLEI)

Dentate nucleus
(fibrous capsule)

*Cerebellar
NEP*

Vestibular nuclear
complex

Prepositus
nucleus

**Fourth
ventricle**

Medullary G/EP

Cochlear nucleus (dorsal)

Inferior cerebellar peduncle

Abducens
nucleus (VI)

**Reticular
formation**

Germinal trigone

Cochlear nucleus (ventral)

Raphe nuclear
complex

Nerve VIII (vestibulocochlear)?

MEDULLA

Dorsal lip of the rhombencephalon
Cochlear neuroepithelium

Spinal tract and nucleus (V)

Facial motor nucleus (VII)

Inferior olive

Anterior extramural migratory stream
(contains pontine gray and reticular tegmental neurons)

Pyramid (*pioneer fibers* of corticospinal tract)

Raphe migration?

PLATE 14A
CR 96 mm, GW 14
Y144-63
Frontal
Section 621

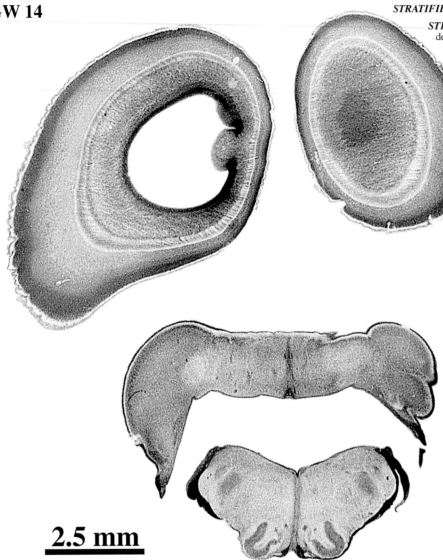

2.5 mm

LAYERS OF THE CORTICAL
STRATIFIED TRANSITIONAL FIELD (STF)

STF1—Superficial fibrous layer with an early developmental stage *(t1)* when many cells are migrating through it, followed by a late stage *(t2)* with sparse cells. Endures as the subcortical white matter.

STF2—Upper cellular layer, the last sojourn zone before cells translocate to the cortical plate.

STF3—Honeycomb trilaminar matrix *(3a, 3b, 3c)* of cells and fibers found only in granular cortices. An extra superficial fibrous layer *(3x)* is prominent in the posterior cortex at GW14.

STF4—Complex middle layer with three developmental stages:
t1– fibrous layer without interspersed cells;
t2– cells and fibers intermingle to form striations;
t3– fibers endure in the deep white matter.

STF5—Deep cellular layer, the first sojourn zone to appear outside the germinal matrix.

STF6—
Late-forming deep layer of callosal fibers outside the germinal matrix.

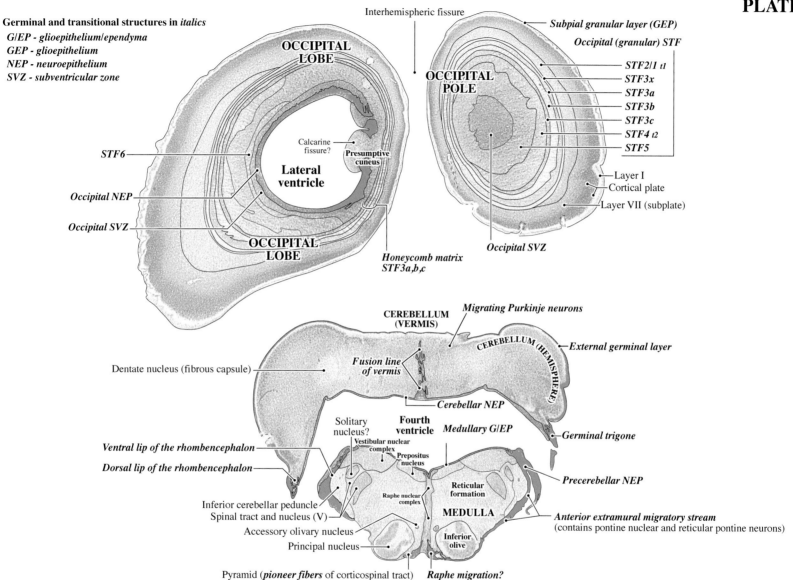

Germinal and transitional structures in *italics*

G/EP - glioepithelium/ependyma
GEP - glioepithelium
NEP - neuroepithelium
SVZ - subventricular zone

Interhemispheric fissure

OCCIPITAL
LOBE

Subpial granular layer (GEP)

Occipital (granular) STF

OCCIPITAL
POLE

STF2/1 t1
STF3x
STF3a
STF3b
STF3c
STF4 t2
STF5

Calcarine
fissure?

Presumptive
cuneus

STF6

Lateral
ventricle

Layer I
Cortical plate
Layer VII (subplate)

Occipital NEP

Occipital SVZ

OCCIPITAL
LOBE

Occipital SVZ

Honeycomb matrix
STF3a,b,c

CEREBELLUM
(VERMIS)

Migrating Purkinje neurons

CEREBELLUM (HEMISPHERE)

External germinal layer

Dentate nucleus (fibrous capsule)

Fusion line
of vermis

Cerebellar NEP

Solitary
nucleus?

Fourth
ventricle

Medullary G/EP

Germinal trigone

Vestibular nuclear
complex

Prepositus
nucleus

Ventral lip of the rhombencephalon

Dorsal lip of the rhombencephalon

Reticular
formation

Precerebellar NEP

Raphe nuclear
complex

MEDULLA

Inferior cerebellar peduncle
Spinal tract and nucleus (V)

Anterior extramural migratory stream
(contains pontine nuclear and reticular pontine neurons)

Accessory olivary nucleus

Principal nucleus

Inferior
olive

Pyramid (*pioneer fibers* of corticospinal tract)

Raphe migration?

PLATE 15A
CR 96 mm, GW 14
Y144-63
Frontal
Section 641

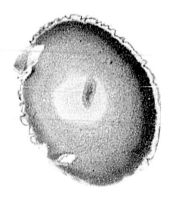

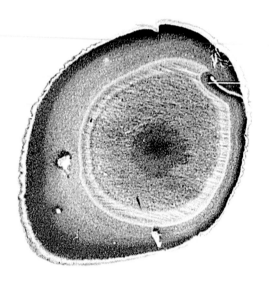

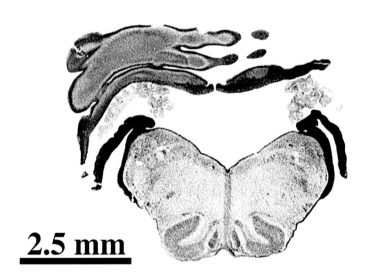

2.5 mm

LAYERS OF THE CORTICAL
STRATIFIED TRANSITIONAL
FIELD (STF)

STF1—Superficial fibrous layer with an early developmental stage *(t1)* when many cells are migrating through it, followed by a late stage *(t2)* with sparse cells. Endures as the subcortical white matter.

STF2—Upper cellular layer, the last sojourn zone before cells translocate to the cortical plate.

STF3—Honeycomb trilaminar matrix *(3a, 3b, 3c)* of cells and fibers found only in granular cortices. An extra superficial fibrous layer *(3x)* is prominent in the posterior cortex at GW14.

STF4—Complex middle layer with three developmental stages:
t1– fibrous layer without interspersed cells;
t2– cells and fibers intermingle to form striations;
t3– fibers endure in the deep white matter.

STF5—Deep cellular layer, the first sojourn zone to appear outside the germinal matrix.

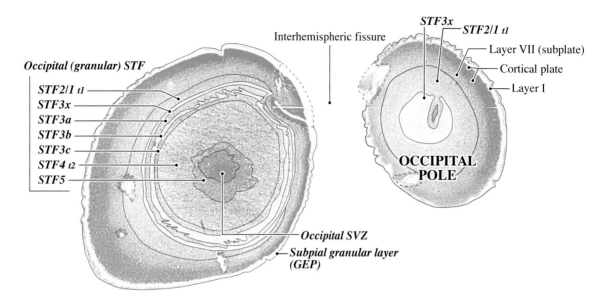

Interhemispheric fissure

Occipital (granular) STF

STF2/1 t1
STF3x
STF3a
STF3b
STF3c
STF4 t2
STF5

STF3x
STF2/1 t1
Layer VII (subplate)
Cortical plate
Layer I

OCCIPITAL POLE

Occipital SVZ
Subpial granular layer (GEP)

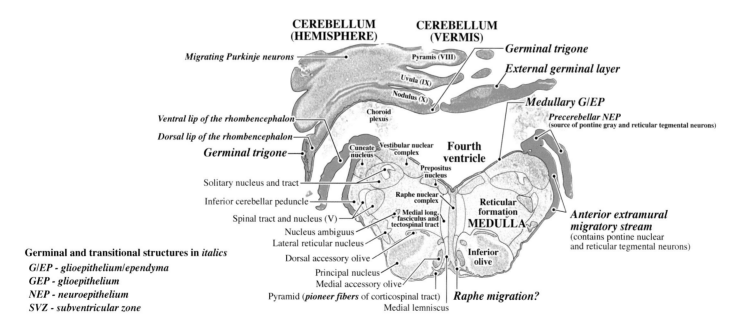

CEREBELLUM (HEMISPHERE)　　**CEREBELLUM (VERMIS)**

Migrating Purkinje neurons

Pyramis (VIII)
Uvula (IX)
Nodulus (X)

Germinal trigone
External germinal layer

Ventral lip of the rhombencephalon
Dorsal lip of the rhombencephalon
Germinal trigone

Choroid plexus

Medullary G/EP
Precerebellar NEP
(source of pontine gray and reticular tegmental neurons)

Cuneate nucleus
Vestibular nuclear complex
Prepositus nucleus

Fourth ventricle

Solitary nucleus and tract
Inferior cerebellar peduncle
Spinal tract and nucleus (V)
Nucleus ambiguus
Lateral reticular nucleus
Dorsal accessory olive
Principal nucleus
Medial accessory olive

Raphe nuclear complex
Medial long. fasciculus and tectospinal tract

Reticular formation MEDULLA

Anterior extramural migratory stream
(contains pontine nuclear and reticular tegmental neurons)

Inferior olive

Pyramid (*pioneer fibers* of corticospinal tract)
Medial lemniscus

Raphe migration?

Germinal and transitional structures in *italics*
 G/EP - *glioepithelium/ependyma*
 GEP - *glioepithelium*
 NEP - *neuroepithelium*
 SVZ - *subventricular zone*

PLATE 16A
CR 96 mm, GW 14
Y144-63
Frontal
Section 661

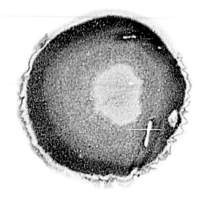

LAYERS OF THE CORTICAL
STRATIFIED TRANSITIONAL
FIELD (STF)

STF1—Superficial fibrous
layer with an early
developmental stage *(t1)*
when many cells are
migrating through it, followed
by a late stage *(t2)* with sparse
cells. Endures as the
subcortical white matter.

STF2—Upper cellular layer,
the last sojourn zone before
cells translocate to the cortical
plate.

STF3—Honeycomb
trilaminar matrix *(3a, 3b, 3c)*
of cells and fibers found only
in granular cortices. An extra
superficial fibrous layer *(3x)*
is prominent in the posterior
cortex at GW14.

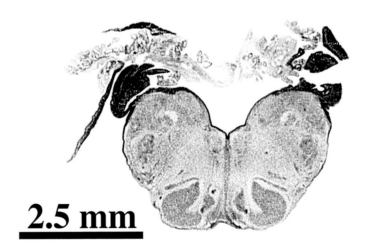

2.5 mm

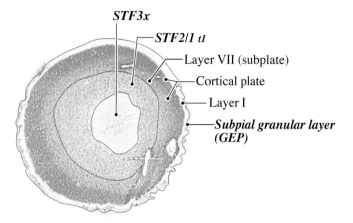

STF3x

STF2/1 *t1*

Layer VII (subplate)

Cortical plate

Layer I

Subpial granular layer (GEP)

LEFT OCCIPITAL POLE

MEDULLA

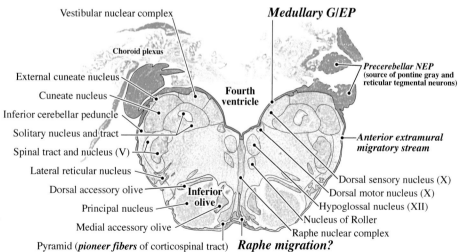

Vestibular nuclear complex

Medullary G/EP

Choroid plexus

Precerebellar NEP
(source of pontine gray and
reticular tegmental neurons)

External cuneate nucleus

Cuneate nucleus

Inferior cerebellar peduncle

Solitary nucleus and tract

Spinal tract and nucleus (V)

Lateral reticular nucleus

Dorsal accessory olive

Principal nucleus

Medial accessory olive

Pyramid (*pioneer fibers* of corticospinal tract)

Fourth ventricle

Anterior extramural migratory stream

Inferior olive

Raphe migration?

Dorsal sensory nucleus (X)

Dorsal motor nucleus (X)

Hypoglossal nucleus (XII)

Nucleus of Roller

Raphe nuclear complex

Germinal and transitional structures in *italics*

G/EP - glioepithelium/ependyma
GEP - glioepithelium
NEP - neuroepithelium

PLATE 17A
CR 96 mm, GW 14
Y144-63
Frontal
Section 321

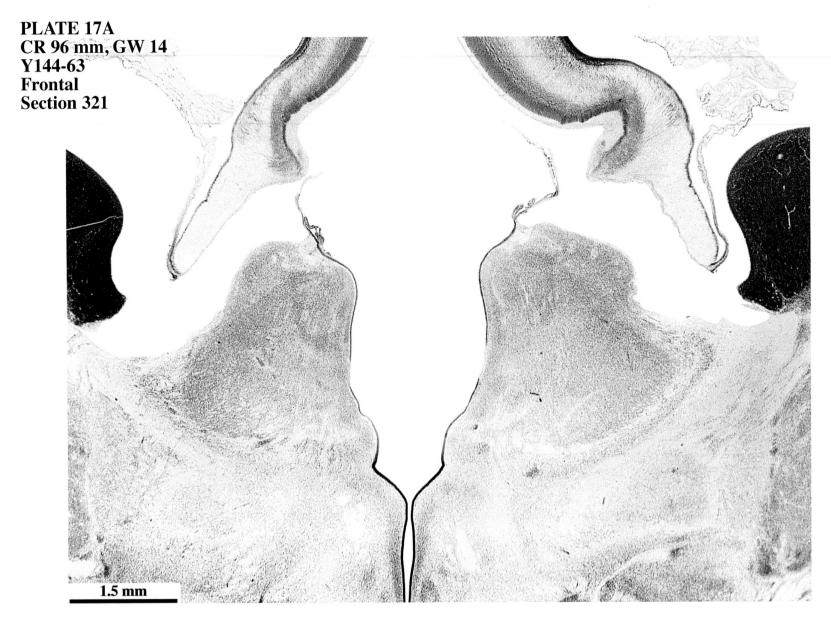

1.5 mm

See the entire Section in Plates 6A and B.

PLATE 17B

Germinal and transitional structures in *italics*

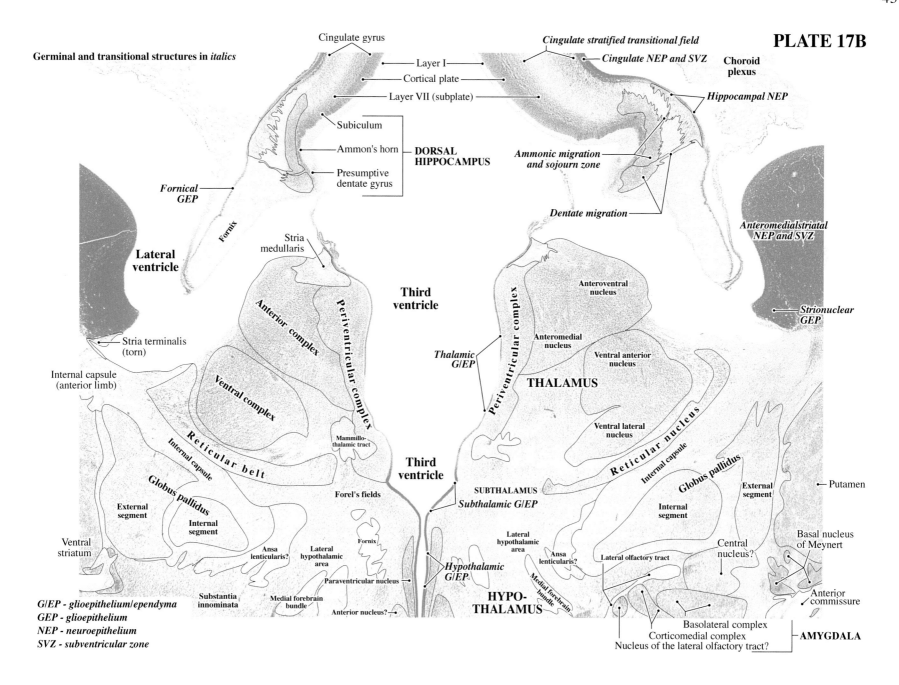

Cingulate gyrus

Cingulate stratified transitional field

Cingulate NEP and SVZ

Choroid plexus

Layer I

Cortical plate

Layer VII (subplate)

Hippocampal NEP

Subiculum

Ammon's horn

DORSAL HIPPOCAMPUS

Ammonic migration and sojourn zone

Presumptive dentate gyrus

Fornical GEP

Dentate migration

Fornix

Anteromedialstriatal NEP and SVZ

Lateral ventricle

Stria medullaris

Anterior complex

Periventricular complex

Third ventricle

Anteroventral nucleus

Strionuclear GEP

Anteromedial nucleus

Stria terminalis (torn)

Periventricular complex

Thalamic G/EP

Ventral anterior nucleus

Internal capsule (anterior limb)

Ventral complex

THALAMUS

Ventral lateral nucleus

Reticular nucleus

Reticular belt

Internal capsule

Mammillo-thalamic tract

Third ventricle

Internal capsule

Globus pallidus

Globus pallidus

External segment

Putamen

Forel's fields

SUBTHALAMUS

Subthalamic G/EP

Internal segment

External segment

Internal segment

Ventral striatum

Ansa lenticularis?

Lateral hypothalamic area

Fornix

Lateral hypothalamic area

Ansa lenticularis?

Central nucleus?

Basal nucleus of Meynert

Hypothalamic G/EP

Lateral olfactory tract

Substantia innominata

Medial forebrain bundle

Paraventricular nucleus

Medial forebrain bundle

HYPO-THALAMUS

Anterior commissure

Anterior nucleus?

Basolateral complex

Corticomedial complex

Nucleus of the lateral olfactory tract?

AMYGDALA

G/EP - glioepithelium/ependyma
GEP - glioepithelium
NEP - neuroepithelium
SVZ - subventricular zone

PLATE 18A
CR 96 mm, GW 14, Y144-63
Frontal
Section 351

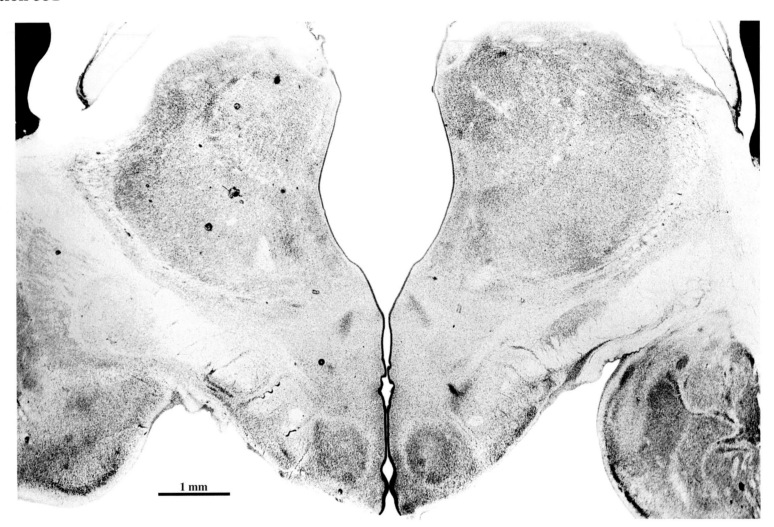

1 mm

See the entire Section 361 in Plates 7A and B.

Germinal and transitional structures in *italics*

G/EP - glioepithelium/ependyma
GEP - glioepithelium
NEP - neuroepithelium
SVZ - subventricular zone

HABENULA
Medial habenula nucleus
Lateral habenula nucleus
Stria medullaris

EPITHALAMUS

Posterior striatal NEP and SVZ
Fornical GEP

Epithalamic G/EP

Dorsolateral nucleus?

Fornix

Lateral ventricle

Periventricular complex

Stria terminalis (torn)

Third ventricle

Dorsomedial nucleus

Lateral geniculate migration?

Optic tract

Internal capsule

Ventral lateral nucleus

Ventral lateral nucleus

Thalamic G/EP

Periventricular complex

THALAMUS

Ventral complex

Reticular belt

Internal capsule

Mammillo-thalamic tract

Reuniens nucleus

Third ventricle

Reticular nucleus

Mammillo-thalamic tract

Zona incerta

Ventral medial nucleus

Ventral medial nucleus

Globus pallidus

SUBTHALAMUS
Forel's fields

Subthalamic G/EP

Subthalamic nucleus

Cerebral peduncle

External segment

Internal segment

Cerebral peduncle

Optic tract

Optic tract

Stria terminalis

Ventral striatum

Ansa lenticularis?

Lateral hypothalamic area

Dorsomedial nucleus

Hypothalamic G/EP

Lateral hypothalamic area

Central nucleus

Central nucleus

Medial nucleus

Optic tract

Medial forebrain bundle

Fornix

Fornix

Medial forebrain bundle

Medial nucleus

Intercalated masses

AMYGDALA

Perifascicular GEP (invades and surrounds fiber tracts)

Optic tract

Ventromedial nucleus

Basal accessory nucleus

Basal nucleus

Cortical nucleus

Primary olfactory cortex/ amygdala junction

Cortical nucleus

Lateral olfactory tract

HYPO-THALAMUS

Lateral olfactory tract

AMYGDALA

Lateral nucleus?

PLATE 19A
CR 96 mm, GW 14, Y144-63
Frontal
Section 401

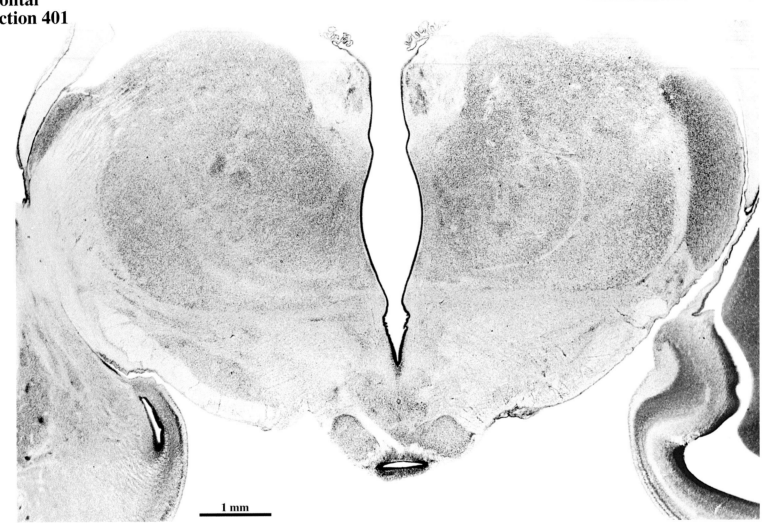

1 mm

See the entire Section 401 in Plates 8A and B.

PLATE 19B

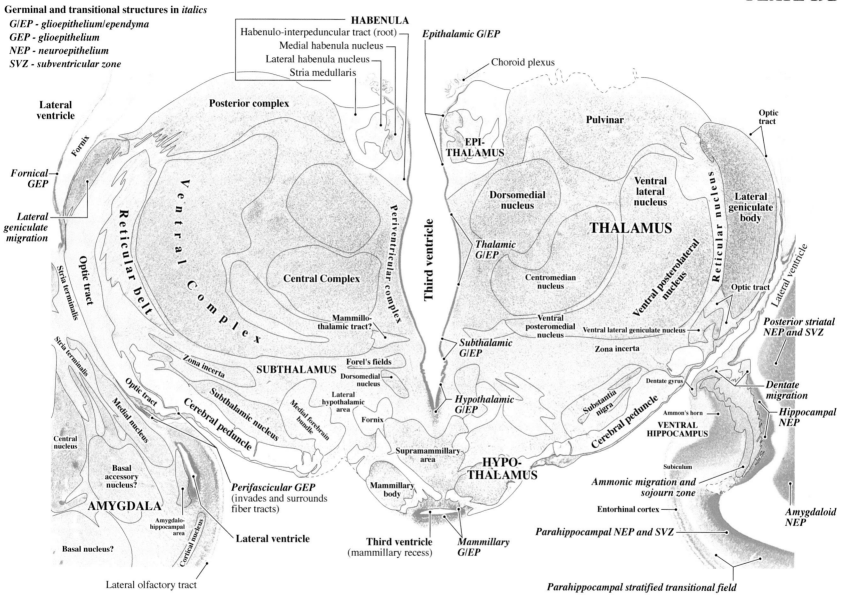

Germinal and transitional structures in *italics*
G/EP - glioepithelium/ependyma
GEP - glioepithelium
NEP - neuroepithelium
SVZ - subventricular zone

HABENULA
Habenulo-interpeduncular tract (root)
Medial habenula nucleus
Lateral habenula nucleus
Stria medullaris

Epithalamic G/EP

Choroid plexus

Lateral ventricle

Posterior complex

Pulvinar

Optic tract

EPI-THALAMUS

Fornical GEP

Fornix

Dorsomedial nucleus

Ventral lateral nucleus

Lateral geniculate body

Lateral geniculate migration

Ventral Complex

Reticular belt

THALAMUS

Reticular nucleus

Lateral ventricle

Optic tract

Stria terminalis

Thalamic G/EP

Central Complex

Periventricular complex

Third ventricle

Centromedian nucleus

Ventral posterolateral nucleus

Optic tract

Stria terminalis

Mammillo-thalamic tract?

Ventral posteromedial nucleus

Ventral lateral geniculate nucleus

Posterior striatal NEP and SVZ

Optic tract

Zona incerta

SUBTHALAMUS

Forel's fields

Subthalamic G/EP

Zona incerta

Medial nucleus

Subthalamic nucleus

Dorsomedial nucleus

Dentate gyrus

Dentate migration

Cerebral peduncle

Lateral hypothalamic area

Hypothalamic G/EP

Substantia nigra

Cerebral peduncle

Ammon's horn

Hippocampal NEP

Central nucleus

Medial forebrain bundle

Fornix

VENTRAL HIPPOCAMPUS

Basal accessory nucleus?

Supramammillary area

HYPO-THALAMUS

Subiculum

Perifascicular GEP (invades and surrounds fiber tracts)

Ammonic migration and sojourn zone

AMYGDALA

Amygdalo-hippocampal area

Cortical nucleus

Mammillary body

Lateral ventricle

Entorhinal cortex

Amygdaloid NEP

Basal nucleus?

Third ventricle (mammillary recess)

Mammillary G/EP

Parahippocampal NEP and SVZ

Lateral olfactory tract

Parahippocampal stratified transitional field

PLATE 20A
CR 96 mm, GW 14, Y144-63
Frontal
Section 421

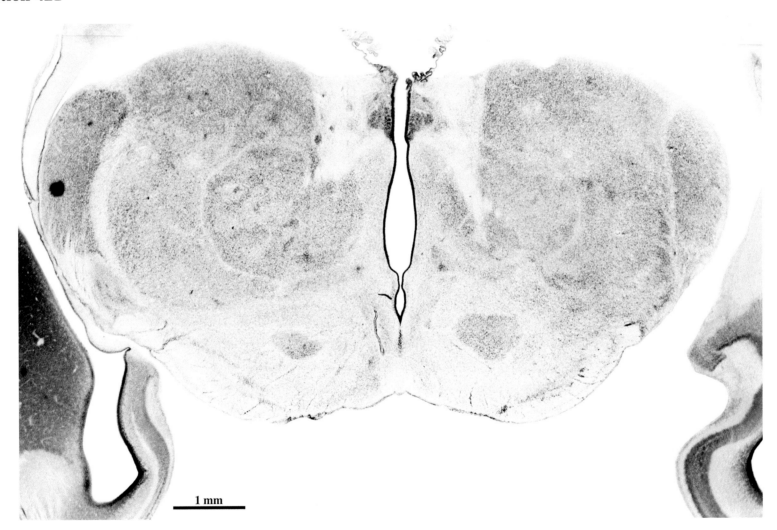

1 mm

See the entire Section 441 in Plates 9A and B.

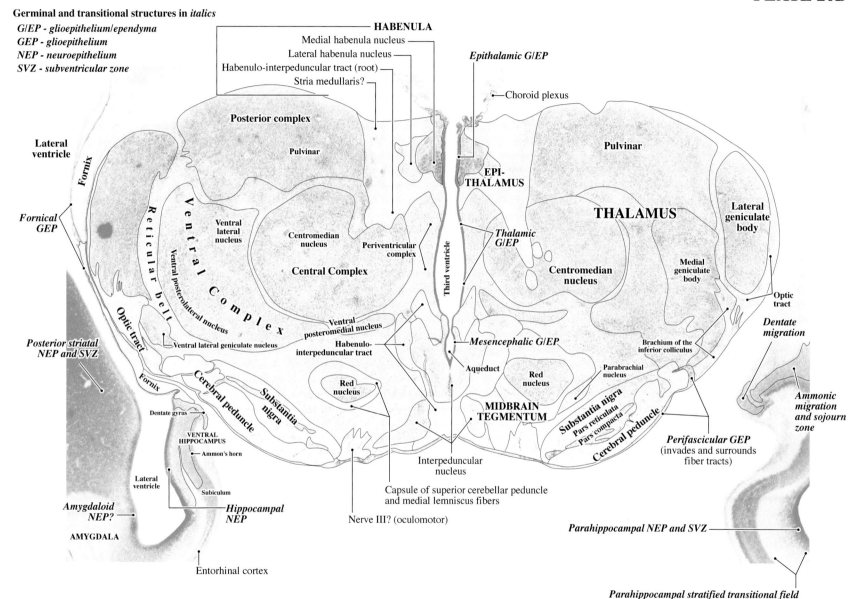

Germinal and transitional structures in *italics*

G/EP - glioepithelium/ependyma
GEP - glioepithelium
NEP - neuroepithelium
SVZ - subventricular zone

50

**PLATE 21
CR 96 mm
GW 14
Y144-63
Frontal
Section 201**

**See the entire Section
in Plates 2A and B.**

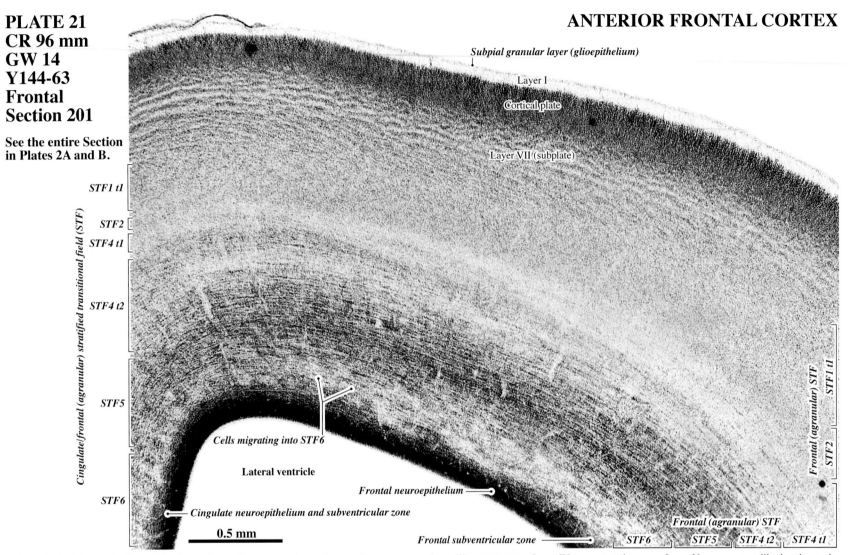

Subpial granular layer (glioepithelium)

Layer I

Cortical plate

Layer VII (subplate)

Cingulate/frontal (agranular) stratified transitional field (STF)

STF1 t1

STF2

STF4 t1

STF4 t2

STF5

Cells migrating into STF6

Lateral ventricle

STF6

Cingulate neuroepithelium and subventricular zone

Frontal neuroepithelium

Frontal subventricular zone

0.5 mm

Frontal (agranular) STF

STF1 t1

STF2

Frontal (agranular) STF

STF6   STF5   STF4 t2   STF4 t1

The **cortical plate** contains Layer VI and some Layer V neurons. Almost all Layer II neurons, most Layer III neurons, many Layer IV neurons, and younger Layer V neurons are still migrating to the cortical plate. Layer VII (the subplate).appears as cohorts of cells horizontally arrayed between fibers forming a striated structure beneath the cortical plate. Many of the cells in the subplate are migrating through it to reach the cortical plate. The subplate blends indistinctly with the ***stratified transitional field (STF).*** In this area of agranular (sparse Layer IV) frontal cortex, five **STF** layers are distinguishable. **STF1** is thick and filled with cells in its early ***t1*** stage. The cells in **STF2** are slightly denser than in **STF1**. **STF4** is the thickest layer with a superficial ***t1*** phase (fibers are there before many cells mingle with them as they migrate to the cortical plate) and a deep ***t2*** phase (cells mingle with the fibers to create striations in the layer). **STF5** is dense and prominent. **STF6** is full of infiltrating callosal fibers intermingled with cells migrating out of the subventricular zone.

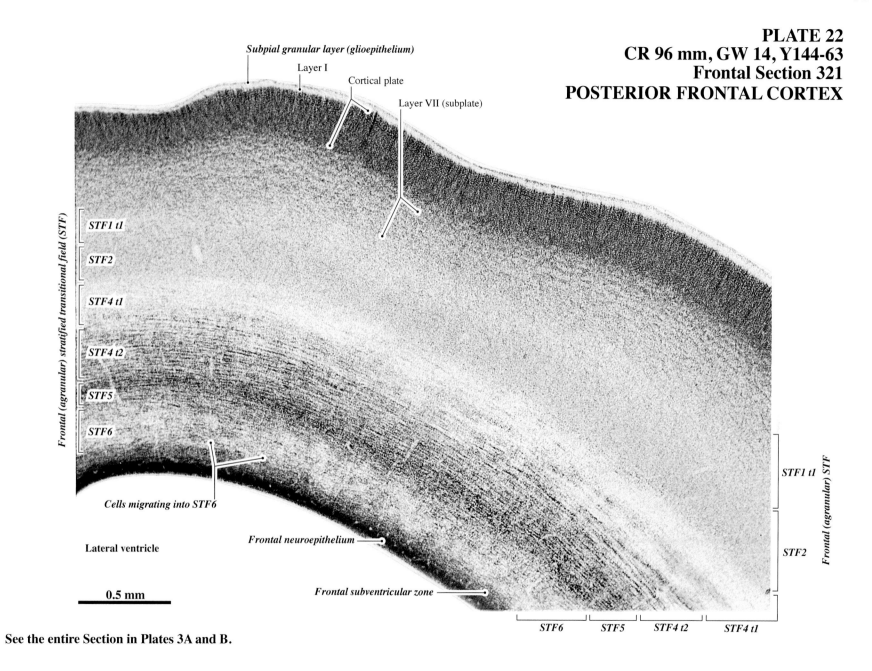

Subpial granular layer (glioepithelium)

Layer I

Cortical plate

Layer VII (subplate)

Frontal (agranular) stratified transitional field (STF)

STF1 t1

STF2

STF4 t1

STF4 t2

STF5

STF6

Cells migrating into STF6

Lateral ventricle

0.5 mm

Frontal neuroepithelium

Frontal subventricular zone

STF1 t1

STF2

Frontal (agranular) STF

STF6   STF5   STF4 t2   STF4 t1

See the entire Section in Plates 3A and B.

PLATE 23
CR 96 mm, GW 14, Y144-63
Frontal Section 521

DORSAL PARIETAL CORTEX

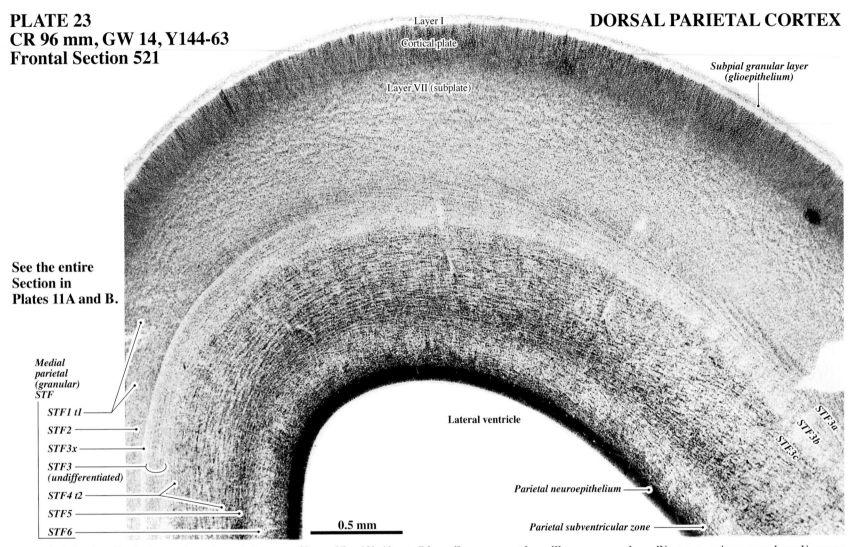

Layer I

Cortical plate

Subpial granular layer
(glioepithelium)

Layer VII (subplate)

**See the entire
Section in
Plates 11A and B.**

Medial
parietal
(granular)
STF

STF1 t1

STF2

STF3x

STF3
(undifferentiated)

STF4 t2

STF5

STF6

Lateral ventricle

STF3a

STF3b

STF3c

Parietal neuroepithelium

Parietal subventricular zone

0.5 mm

The **cortical plate** is uniformly dense and contains settled neurons of Layers VI and V. Almost all Layer II neurons, most Layer III neurons, many Layer IV neurons, and even some Layer V neurons are still migrating to the cortical plate. Layer VII (the subplate) has a striated structure beneath the cortical plate, similar to the subplate in the frontal cortex (**Plates 21** and **22**), that blends indistinctly with the *stratified transitional field (STF)*. The *STF* has all six layers characteristic of granular cortices where Layer IV is prominent. *STF3x* is a distinct fibrous layer in the medial parietal cortex and the far-lateral parietal cortex (**Plate 22**); it is only found at this stage of development. The remaining *STF3* is subdivided into *a, b, and c* sublayers laterally but is undifferentiated medially. *STF4* is in its *t2* developmental stage when cells migrating to the cortical plate intermingle with fibers that have grown in from the posterior limb of the internal capsule. *STF5* forms a dense and distinct band throughout the dorsal parietal cortex. *STF6* contains *uncrossed* (ipsilateral) callosal fibers interspersed with cells migraing through it on their way to the cortical plate.

LATERAL PARIETAL CORTEX

PLATE 24
CR 96 mm, GW 14, Y144-63, Frontal Section 521

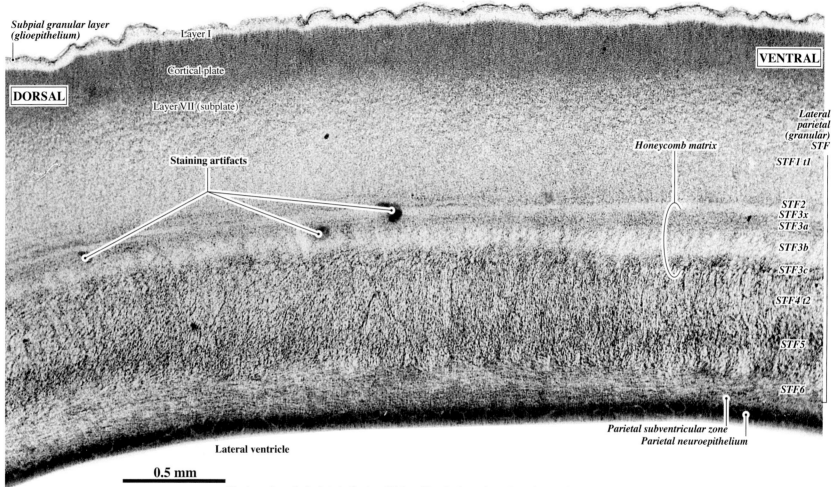

Subpial granular layer (glioepithelium)

Layer I

Cortical plate

VENTRAL

DORSAL

Layer VII (subplate)

Lateral parietal (granular) STF

STF1 t1

Staining artifacts

Honeycomb matrix

STF2
STF3x
STF3a

STF3b

STF3c

STF4 t2

STF5

STF6

Parietal subventricular zone
Parietal neuroepithelium

Lateral ventricle

0.5 mm

See the entire Section in Plates 11A and B.

The lateral **cortical plate** in Section 521 is uniformly dense throughout, just as the dorsal cortical plate (**Plate 23**). The *stratified transitional field (STF)* is the thickest part of the developing cortex and has the "full flowering" of all six layers characteristic of granular cortices where Layer IV is prominent. *STF1* is in its *t1* phase when it is filled with cells migrating to the cortical plate. *STF2* is distinguishable as a denser cellular band at the base of *STF1*. *STF3* is well differentiated into a superficial cellular layer *(a)*, an intermediate fibrous layer *(b)* with streaks of radially aligned migrating cells *(honeycomb matrix)*, and a deep cellular layer *(c)* where cells accumulate before migrating through *STF3b*. In addition, there is a distinct fibrous layer *(STF3x)* that clearly delineates *STF2* and *STF3*; *STF3x* is only found at this stage of development in the posterior cortex. *STF4*, *STF5*, and *STF6* are similar to dorsal parietal cortex (**Plate 23**), except there are fewer callosal fibers in *STF6*.

PART III: Y37-63
CR 145 mm (GW 17)
Sagittal

This specimen is a stillborn fetus of undetermined sex (Yakovlev case number BR W-37-63; referred to here as Y37-63) with a crown-rump length (CR) of 145 mm estimated to be at gestational week (GW) 17. The brain was cut in the sagittal plane in 414 35-μm thick sections and is classified as a Normative Control in the Yakovlev Collection (Haleem, 1990). Since there is no photograph of this brain before it was embedded and cut, we turned to the comprehensive atlas that Retzius published in 1896 showing whole fetal brains in medial, lateral, superior, and inferior views and midline sagittally cut brains. **Figure 8**, taken from Retzius (1896), shows the midline sagittal surface of a brain from a specimen that is at the same age as Y37-63. Low-magnification photographs of six Nissl-stained sections from the right hemisphere are shown in **Plates 25–30**. High-magnification views of different regions of the cerebral cortex are shown in **Plates 31** and **32**, and of the midline cerebellar cortex in **Plates 33** and **34**.

In the telencephalon, the ***cortical neuroepithelium/subventricular zone*** is bordered by sharply defined layers of the ***stratified transitional fields (STF)*** in all lobes of the cerebral cortex. Lateral sagittal sections show the differences in the ***STF*** between future sensory and future motor areas. The high-magnification cortical images show sharp delineation of the various ***STF*** layers. This specimen has an especially well-preserved *rostral migratory stream* extending into the olfactory bulb and peduncle. The ***ammonic migration*** is more definite in the hippocampus, and the ***dentate migration*** can be traced from its ventricular source near the fimbria/fornix. The ***subgranular zone*** fills up most of the hilus and is just beginning to generate dentate granule cells. There is a definite dentate granular layer. A massive ***neuroepithelium/subventricular zone*** in the nucleus accumbens and striatum is the site of neurogenesis (and gliogenesis).

For the most part, brainstem structures appear more mature than in the telencephalon, but lateral sections of the thalamus show a darkly staining stream of migrating cells heading toward the lateral geniculate body. In the medulla, there is a remnant of the ***precerebellar neuroepithelium***. The ***anterior extramural migratory stream*** enters the posterior pontine gray. The cerebellar peduncle is enlarging, as is the transpontine corticofugal tract. However, the pyramids at the base of the medulla are small, awaiting the ingrowth of cortical axons.

The cerebellum in this specimen shows primary lobules in the midline vermal cortex, and deep nuclei are settled in the core of white matter. The entire surface of the cerebellar cortex is covered by the prominent ***external germinal layer (egl)*** that is actively producing basket, stellate, and granule cells. Lamination in the cortex is immature, except for a thin molecular layer beneath the ***egl***. Most cortical neurons, including Purkinje cells, are still migrating. Purkinje cells appear to arrive at the cortical surface in clumps, as shown in the high-magnification views of the declive (**Plate 34**). Sublobulation has barely begun in the vermis and is even less obvious the hemispheres. The ***germinal trigone*** is large at the base of the nodulus and along the floccular peduncle.

GW17 MIDLINE SAGITTAL VIEW

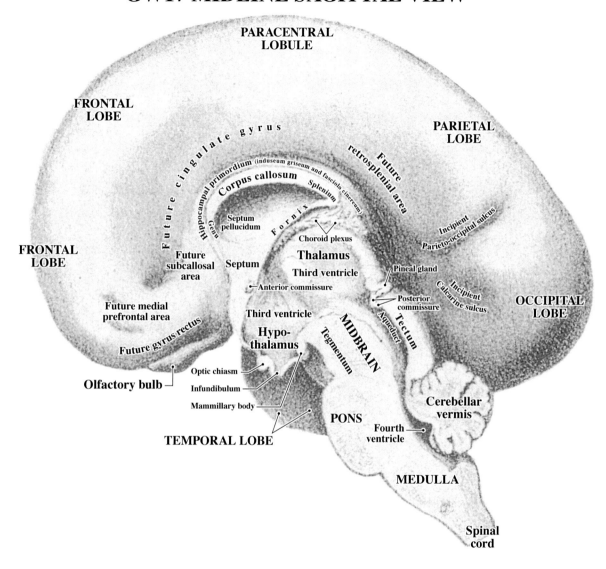

Figure 8. Midline sagittal view of a GW17 brain with major structures in the cerebral hemispheres and brainstem labeled. (This is Figure 17 in Table 4, Volume 2, Retzius, 1896.)

PLATE 25A
CR 145 mm, GW 17
Y37-63
Sagittal
Section 61

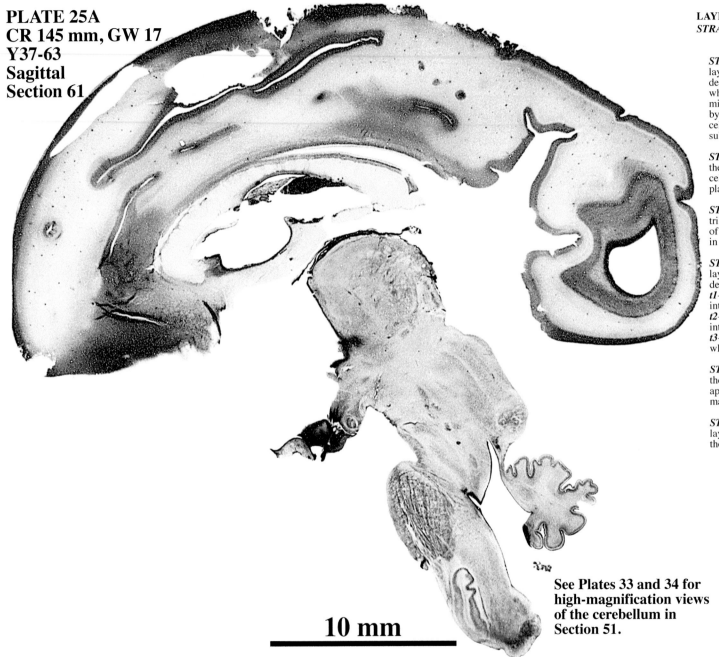

10 mm

See Plates 33 and 34 for
high-magnification views
of the cerebellum in
Section 51.

LAYERS OF THE CORTICAL
STRATIFIED TRANSITIONAL
FIELD (STF)

STF1—Superficial fibrous layer with an early developmental stage *(t1)* when many cells are migrating through it, followed by a late stage *(t2)* with sparse cells. Endures as the subcortical white matter.

STF2—Upper cellular layer, the last sojourn zone before cells translocate to the cortical plate.

STF3—Honeycomb trilaminar matrix *(3a, 3b, 3c)* of cells and fibers found only in granular cortices.

STF4—Complex middle layer with three developmental stages:
t1– fibrous layer without interspersed cells;
t2– cells and fibers intermingle to form striations;
t3– fibers endure in the deep white matter.

STF5—Deep cellular layer, the first sojourn zone to appear outside the germinal matrix.

STF6—Late-forming deep layer of callosal fibers outside the germinal matrix.

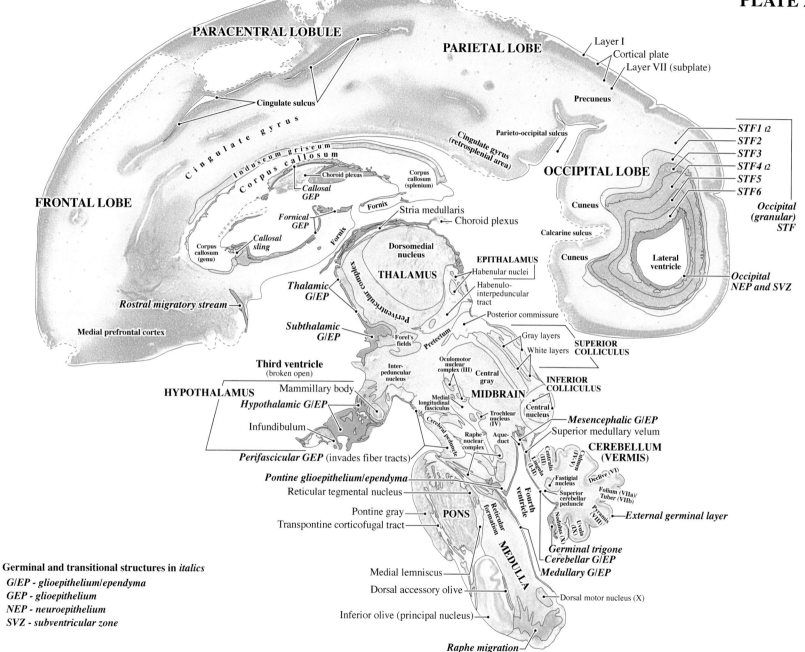

PARACENTRAL LOBULE

PARIETAL LOBE

Layer I
Cortical plate
Layer VII (subplate)

Cingulate sulcus

Precuneus

Cingulate gyrus

Parieto-occipital sulcus

Cingulate gyrus (retrosplenial area)

STF1 t2
STF2
STF3
STF4 t2
STF5
STF6

Induseum griseum

Corpus callosum

Choroid plexus

Corpus callosum (splenium)

OCCIPITAL LOBE

Occipital (granular) STF

FRONTAL LOBE

Callosal GEP

Fornix

Stria medullaris

Cuneus

Corpus callosum

Fornical GEP

Fornix

Choroid plexus

Calcarine sulcus

Corpus callosum (genu)

Callosal sling

Dorsomedial nucleus

THALAMUS

EPITHALAMUS
Habenular nuclei

Cuneus

Lateral ventricle

Occipital NEP and SVZ

Rostral migratory stream

Thalamic G/EP

Periventricular complex

Habenulo-interpeduncular tract

Posterior commissure

Medial prefrontal cortex

Subthalamic G/EP

Forel's fields

Pretectum

Gray layers
White layers

SUPERIOR COLLICULUS

Third ventricle
(broken open)

Inter-peduncular nucleus

Oculomotor nuclear complex (III)

Central gray

INFERIOR COLLICULUS

HYPOTHALAMUS

Mammillary body

Medial longitudinal fasciculus

MIDBRAIN

Central nucleus

Mesencephalic G/EP

Hypothalamic G/EP

Trochlear nucleus (IV)

Aqueduct

Superior medullary velum

Infundibulum

Cerebral peduncle

Raphe nuclear complex

Centralis (III)
Lingula (I-II)
Culmen (IV-V)

CEREBELLUM (VERMIS)

Perifascicular GEP (invades fiber tracts)

Pontine glioepithelium/ependyma

Fastigial nucleus

Declive (VI)

Reticular tegmental nucleus

Superior cerebellar peduncle

Folium (VIIa)/Tuber (VIIb)

Pontine gray

PONS

Reticular formation

Fourth ventricle

Pyramis (VIII)

External germinal layer

Transpontine corticofugal tract

Nodulus (X)

Uvula (IX)

MEDULLA

Germinal trigone
Cerebellar G/EP

Medullary G/EP

Medial lemniscus

Dorsal accessory olive

Dorsal motor nucleus (X)

Inferior olive (principal nucleus)

Raphe migration

Germinal and transitional structures in *italics*

G/EP - glioepithelium/ependyma
GEP - glioepithelium
NEP - neuroepithelium
SVZ - subventricular zone

58

PLATE 26A
CR 145 mm, GW 17
Y37-63
Sagittal
Section 111

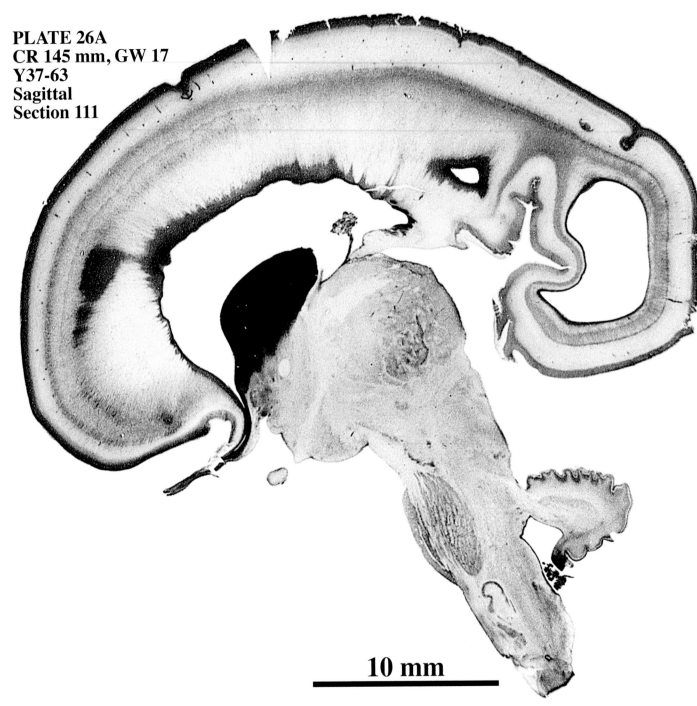

LAYERS OF THE CORTICAL
STRATIFIED TRANSITIONAL
FIELD (STF)

STF1—Superficial fibrous layer with an early developmental stage *(t1)* when many cells are migrating through it, followed by a late stage *(t2)* with sparse cells. Endures as the subcortical white matter.

STF2—Upper cellular layer, the last sojourn zone before cells translocate to the cortical plate.

STF3—Honeycomb trilaminar matrix *(3a, 3b, 3c)* of cells and fibers found only in granular cortices.

STF4—Complex middle layer with three developmental stages:
t1– fibrous layer without interspersed cells;
t2– cells and fibers intermingle to form striations;
t3– fibers endure in the deep white matter.

STF5—Deep cellular layer, the first sojourn zone to appear outside the germinal matrix.

STF6—Late-forming deep layer of callosal fibers outside the germinal matrix.

10 mm

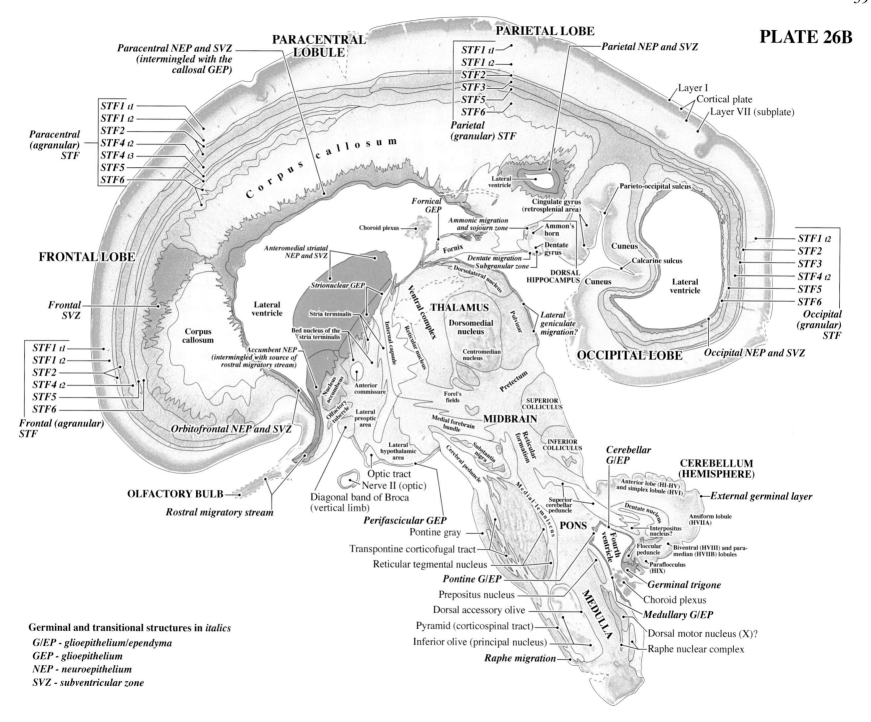

PLATE 26B

Paracentral NEP and SVZ (intermingled with the callosal GEP)

PARACENTRAL LOBULE

PARIETAL LOBE

Parietal NEP and SVZ

STF1 t1
STF1 t2
STF2
STF3
STF5
STF6

Layer I
Cortical plate
Layer VII (subplate)

STF1 t1
STF1 t2
STF2
STF4 t2
STF4 t3
STF5
STF6

Paracentral (agranular) STF

Parietal (granular) STF

C o r p u s   c a l l o s u m

Lateral ventricle

Parieto-occipital sulcus

Fornical GEP

Cingulate gyrus (retrosplenial area)

Choroid plexus

Ammonic migration and sojourn zone

Ammon's horn

Cuneus

FRONTAL LOBE

Anteromedial striatal NEP and SVZ

Fornix

Dentate gyrus

Dentate migration

Calcarine sulcus

Subgranular zone

DORSAL HIPPOCAMPUS

Cuneus

Lateral ventricle

STF1 t2
STF2
STF3
STF4 t2
STF5
STF6

Strionuclear GEP

Dorsolateral nucleus

Frontal SVZ

Lateral ventricle

Stria terminalis

Ventral complex

THALAMUS

Pulvinar

Lateral geniculate migration?

Occipital (granular) STF

Bed nucleus of the stria terminalis

Corpus callosum

Dorsomedial nucleus

Internal capsule

Reticular nucleus

OCCIPITAL LOBE

Occipital NEP and SVZ

STF1 t1
STF1 t2
STF2
STF4 t2
STF5
STF6

Accumbent NEP (intermingled with source of rostral migratory stream)

Centromedian nucleus

Pretectum

Frontal (agranular) STF

Nucleus accumbens

Anterior commissure

Forel's fields

SUPERIOR COLLICULUS

Orbitofrontal NEP and SVZ

Olfactory tubercle

Lateral preoptic area

MIDBRAIN

Substantia nigra

Reticular formation

INFERIOR COLLICULUS

Cerebellar G/EP

CEREBELLUM (HEMISPHERE)

Medial forebrain bundle

Lateral hypothalamic area

Cerebral peduncle

Anterior lobe (HI-HV) and simplex lobule (HVI)

External germinal layer

OLFACTORY BULB

Optic tract

Nerve II (optic)

Diagonal band of Broca (vertical limb)

Medial lemniscus

Dentate nucleus

Ansiform lobule (HVIIA)

Rostral migratory stream

Perifascicular GEP

Superior cerebellar peduncle

Interpositus nucleus?

Biventral (HVIII) and paramedian (HVIIB) lobules

Pontine gray

Flocculor peduncle

Paraflocculus (HIX)

Transpontine corticofugal tract

PONS

Fourth ventricle

Germinal trigone

Reticular tegmental nucleus

Choroid plexus

Pontine G/EP

MEDULLA

Medullary G/EP

Prepositus nucleus

Dorsal accessory olive

Dorsal motor nucleus (X)?

Pyramid (corticospinal tract)

Raphe nuclear complex

Inferior olive (principal nucleus)

Raphe migration

Germinal and transitional structures in *italics*

G/EP - glioepithelium/ependyma
GEP - glioepithelium
NEP - neuroepithelium
SVZ - subventricular zone

60

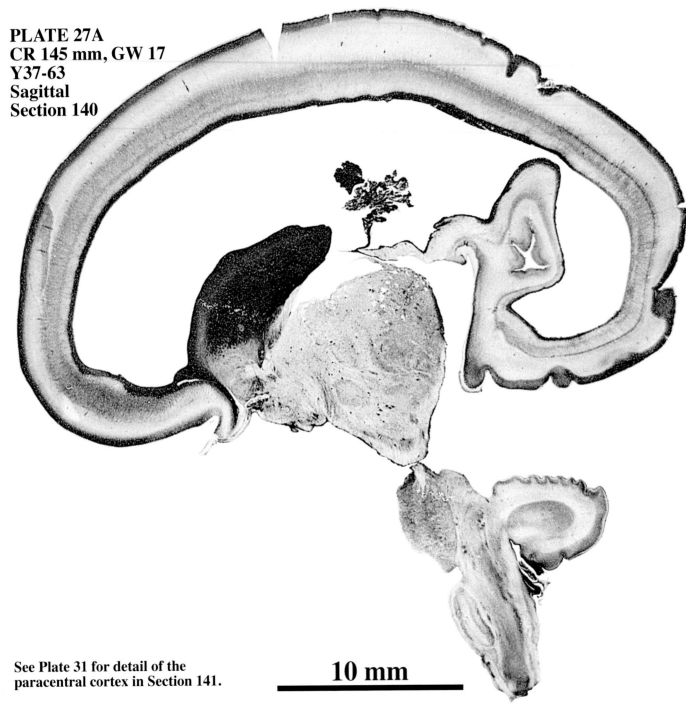

LAYERS OF THE CORTICAL
*STRATIFIED TRANSITIONAL
FIELD (STF)*

STF1—Superficial fibrous
layer with an early
developmental stage *(t1)*
when many cells are
migrating through it, followed
by a late stage *(t2)* with sparse
cells. Endures as the
subcortical white matter.

STF2—Upper cellular layer,
the last sojourn zone before
cells translocate to the cortical
plate.

STF3—Honeycomb
trilaminar matrix *(3a, 3b, 3c)*
of cells and fibers found only
in granular cortices.

STF4—Complex middle
layer with three
developmental stages:
t1– fibrous layer without
interspersed cells;
t2– cells and fibers
intermingle to form striations;
t3– fibers endure in the deep
white matter.

STF5—Deep cellular layer,
the first sojourn zone to
appear outside the germinal
matrix.

STF6—Late-forming deep
layer of callosal fibers outside
the germinal matrix.

**See Plate 31 for detail of the
paracentral cortex in Section 141.**

10 mm

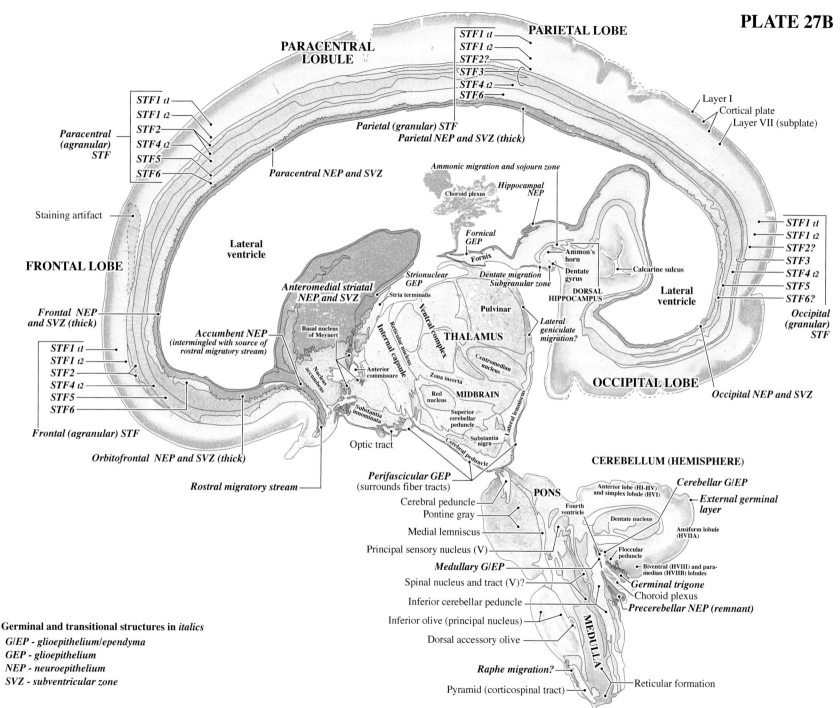

PARACENTRAL
LOBULE

PARIETAL LOBE

STF1 *t1*
STF1 *t2*
STF2?
STF3
STF4 *t2*
STF6

Layer I
Cortical plate
Layer VII (subplate)

STF1 *t1*
STF1 *t2*
STF2
STF4 *t2*
STF5
STF6

*Paracentral
(agranular)
STF*

Paracentral NEP and SVZ

Parietal (granular) STF
Parietal NEP and SVZ (thick)

Staining artifact

FRONTAL LOBE

Lateral
ventricle

Ammonic migration and sojourn zone

Choroid plexus

*Hippocampal
NEP*

STF1 *t1*
STF1 *t2*
STF2?
STF3
STF4 *t2*
STF5
STF6?

*Fornical
GEP*

Fornix

Ammon's
horn

Calcarine sulcus

Lateral
ventricle

*Occipital
(granular)
STF*

*Frontal NEP
and SVZ (thick)*

*Strionuclear
GEP*

*Anteromedial striatal
NEP and SVZ*

Stria terminalis

Dentate migration

Dentate
gyrus

Subgranular zone

DORSAL
HIPPOCAMPUS

*Lateral
geniculate
migration?*

Basal nucleus
of Meynert

Pulvinar

*Accumbent NEP
(intermingled with source of
rostral migratory stream)*

Ventral
complex

Internal capsule

Reticular nucleus

THALAMUS

STF1 *t1*
STF1 *t2*
STF2
STF4 *t2*
STF5
STF6

Nucleus
accumbens

Anterior
commissure

Centromedian
nucleus

Occipital NEP and SVZ

OCCIPITAL LOBE

Zona incerta

Red
nucleus

MIDBRAIN

Substantia
innominata

Frontal (agranular) STF

Superior
cerebellar
peduncle

Substantia
nigra

Cerebral peduncle

Lateral lemniscus

Orbitofrontal NEP and SVZ (thick)

Optic tract

Perifascicular GEP
(surrounds fiber tracts)

CEREBELLUM (HEMISPHERE)

Rostral migratory stream

Cerebral peduncle

PONS

Anterior lobe (HI-HV)
and simplex lobule (HVI)

Cerebellar G/EP

*External germinal
layer*

Pontine gray

Fourth
ventricle

Dentate nucleus

Ansiform lobule
(HVIIA)

Medial lemniscus

Principal sensory nucleus (V)

Floccular
peduncle

Biventral (HVIII) and paramedian (HVIIB) lobules

Medullary G/EP

Germinal trigone

Spinal nucleus and tract (V)?

Choroid plexus

Inferior cerebellar peduncle

MEDULLA

Precerebellar NEP (remnant)

Inferior olive (principal nucleus)

Dorsal accessory olive

Germinal and transitional structures in *italics*

G/EP - glioepithelium/ependyma
GEP - glioepithelium
NEP - neuroepithelium
SVZ - subventricular zone

Raphe migration?

Pyramid (corticospinal tract)

Reticular formation

PLATE 28A
CR 145 mm, GW 17
Y37-63
Sagittal
Section 160

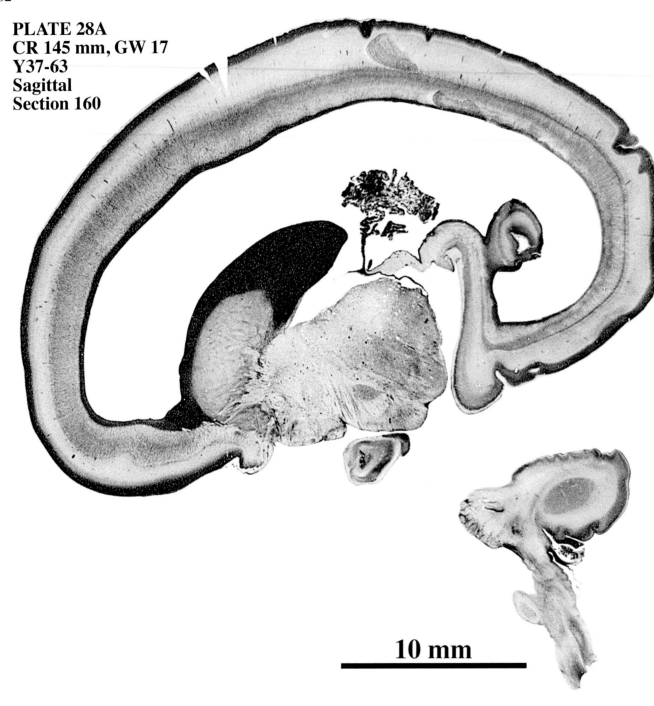

10 mm

LAYERS OF THE CORTICAL
STRATIFIED TRANSITIONAL
FIELD (STF)

STF1—Superficial fibrous layer with an early developmental stage *(t1)* when many cells are migrating through it, followed by a late stage *(t2)* with sparse cells. Endures as the subcortical white matter.

STF2—Upper cellular layer, the last sojourn zone before cells translocate to the cortical plate.

STF3—Honeycomb trilaminar matrix *(3a, 3b, 3c)* of cells and fibers found only in granular cortices.

STF4—Complex middle layer with three developmental stages:
t1– fibrous layer without interspersed cells;
t2– cells and fibers intermingle to form striations;
t3– fibers endure in the deep white matter.

STF5—Deep cellular layer, the first sojourn zone to appear outside the germinal matrix.

STF6—Late-forming deep layer of callosal fibers outside the germinal matrix.

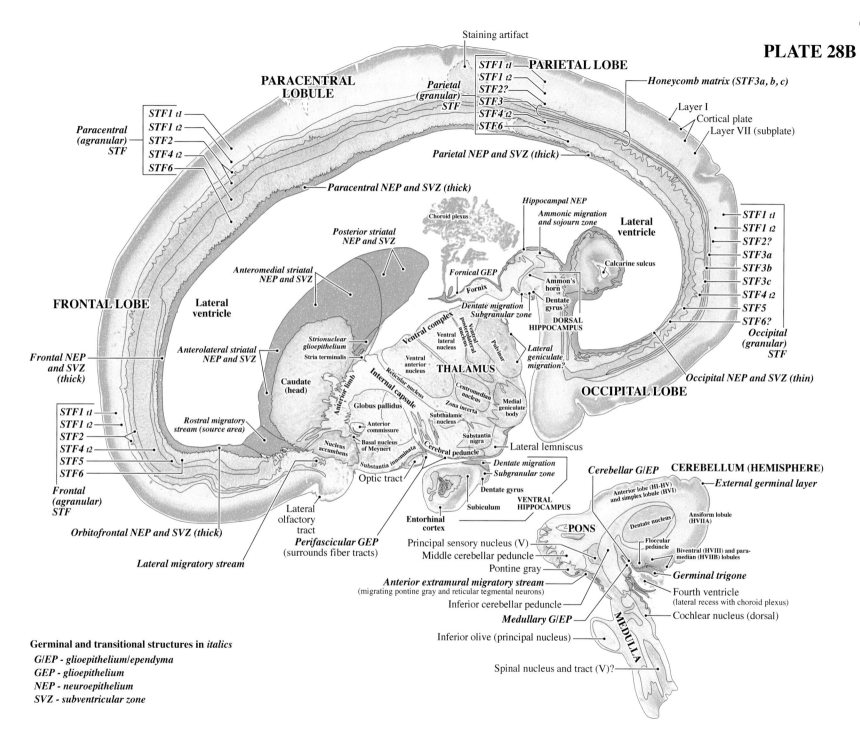

Staining artifact

**PARACENTRAL
LOBULE**

PARIETAL LOBE

STF1 t1
STF1 t2
STF2?
STF3
STF4 t2
STF6

*Parietal
(granular)
STF*

Honeycomb matrix (STF3a, b, c)

Layer I
Cortical plate
Layer VII (subplate)

STF1 t1
STF1 t2
STF2
STF4 t2
STF6

*Paracentral
(agranular)
STF*

Parietal NEP and SVZ (thick)

Paracentral NEP and SVZ (thick)

Hippocampal NEP

Choroid plexus

*Ammonic migration
and sojourn zone*

**Lateral
ventricle**

Calcarine sulcus

STF1 t1
STF1 t2
STF2?
STF3a
STF3b
STF3c
STF4 t2
STF5
STF6?

*Occipital
(granular)
STF*

*Posterior striatal
NEP and SVZ*

Fornical GEP

Fornix

Ammon's
horn

Dentate
gyrus

**DORSAL
HIPPOCAMPUS**

*Anteromedial striatal
NEP and SVZ*

*Dentate migration
Subgranular zone*

FRONTAL LOBE

**Lateral
ventricle**

Ventral complex

Ventral
lateral
nucleus

Ventral
posterolateral
nucleus

Pulvinar

*Lateral
geniculate
migration?*

OCCIPITAL LOBE

*Frontal NEP
and SVZ
(thick)*

*Anterolateral striatal
NEP and SVZ*

*Strionuclear
glioepithelium*

Stria terminalis

Ventral
anterior
nucleus

THALAMUS

Reticular nucleus

Centromedian
nucleus

Zona incerta

Medial
geniculate
body

Occipital NEP and SVZ (thin)

**Caudate
(head)**

Globus pallidus

Subthalamic
nucleus

STF1 t1
STF1 t2
STF2
STF4 t2
STF5
STF6

*Frontal
(agranular)
STF*

*Rostral migratory
stream (source area)*

Anterior
commissure

Internal capsule

Anterior limb

Substantia
nigra

Lateral lemniscus

Nucleus
accumbens

Basal nucleus
of Meynert

Substantia innominata

Cerebral peduncle

*Dentate migration
Subgranular zone*

Cerebellar G/EP

CEREBELLUM (HEMISPHERE)

Optic tract

Dentate gyrus

**VENTRAL
HIPPOCAMPUS**

Anterior lobe (HI-HV)
and simplex lobule (HVI)

External germinal layer

Orbitofrontal NEP and SVZ (thick)

Subiculum

Ansiform lobule
(HVIIA)

Lateral
olfactory
tract

**Entorhinal
cortex**

Dentate nucleus

Lateral migratory stream

Principal sensory nucleus (V)

PONS

Floccular
peduncle

Biventral (HVIII) and paramedian (HVIIB) lobules

Perifascicular GEP
(surrounds fiber tracts)

Middle cerebellar peduncle

Pontine gray

Germinal trigone

Anterior extramural migratory stream
(migrating pontine gray and reticular tegmental neurons)

Inferior cerebellar peduncle

Fourth ventricle
(lateral recess with choroid plexus)

Cochlear nucleus (dorsal)

Medullary G/EP

MEDULLA

Inferior olive (principal nucleus)

Spinal nucleus and tract (V)?

Germinal and transitional structures in *italics*

G/EP - glioepithelium/ependyma
GEP - glioepithelium
NEP - neuroepithelium
SVZ - subventricular zone

PLATE 29A
CR 145 mm, GW 17
Y37-63
Sagittal
Section 201

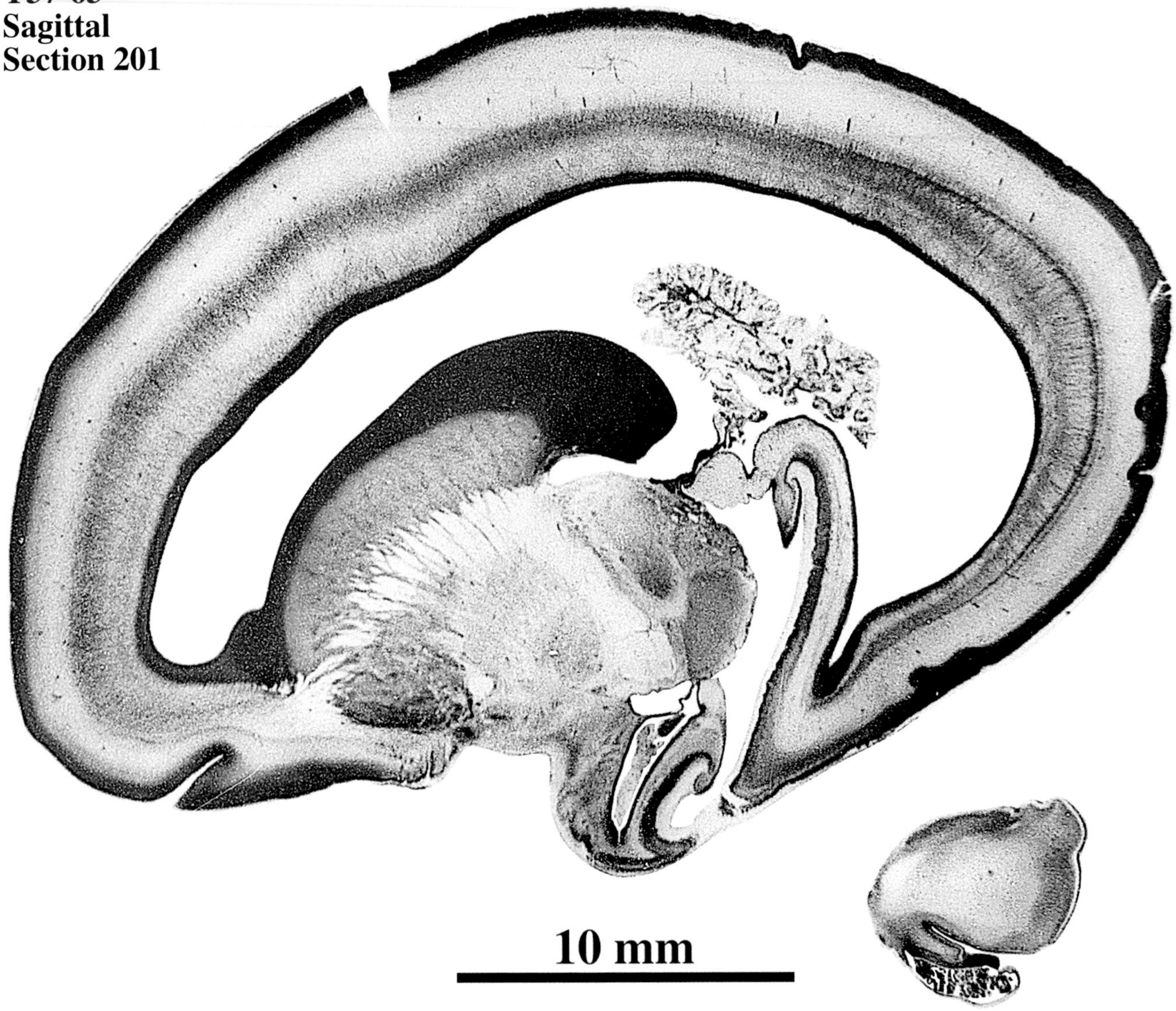

10 mm

LAYERS OF THE CORTICAL
STRATIFIED TRANSITIONAL
FIELD (STF)

STF1—Superficial fibrous layer with an early developmental stage *(t1)* when many cells are migrating through it, followed by a late stage *(t2)* with sparse cells. Endures as the subcortical white matter.

STF2—Upper cellular layer, the last sojourn zone before cells translocate to the cortical plate.

STF3—Honeycomb trilaminar matrix *(3a, 3b, 3c)* of cells and fibers found only in granular cortices.

STF4—Complex middle layer with three developmental stages:
t1– fibrous layer without interspersed cells;
t2– cells and fibers intermingle to form striations;
t3– fibers endure in the deep white matter.

STF5—Deep cellular layer, the first sojourn zone to appear outside the germinal matrix.

STF6—Late-forming deep layer of callosal fibers outside the germinal matrix.

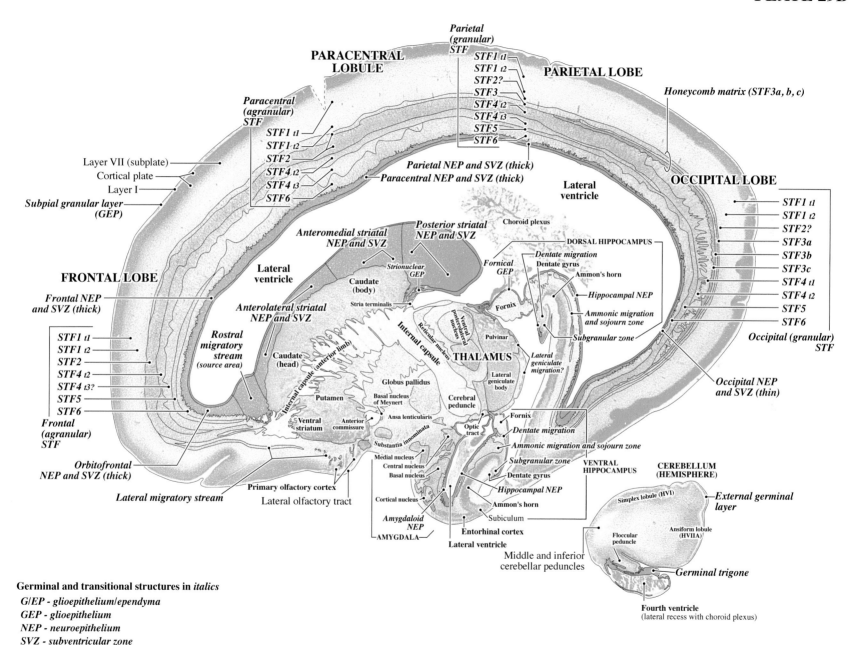

Parietal (granular) STF

PARACENTRAL LOBULE

STF1 t1
STF1 t2
STF2?
STF3
STF4 t2
STF4 t3
STF5
STF6

PARIETAL LOBE

Honeycomb matrix (STF3a, b, c)

Paracentral (agranular) STF

STF1 t1
STF1 t2
STF2
STF4 t2
STF4 t3
STF6

Layer VII (subplate)
Cortical plate
Layer I

Subpial granular layer (GEP)

Parietal NEP and SVZ (thick)
Paracentral NEP and SVZ (thick)

Lateral ventricle

OCCIPITAL LOBE

STF1 t1
STF1 t2
STF2?
STF3a
STF3b
STF3c
STF4 t1
STF4 t2
STF5
STF6

Occipital (granular) STF

Anteromedial striatal NEP and SVZ

Posterior striatal NEP and SVZ

Choroid plexus

DORSAL HIPPOCAMPUS

Dentate migration
Dentate gyrus
Ammon's horn

FRONTAL LOBE

Lateral ventricle

Strionuclear GEP

Caudate (body)

Fornical GEP

Hippocampal NEP

Frontal NEP and SVZ (thick)

Anterolateral striatal NEP and SVZ

Stria terminalis

Fornix

Ammonic migration and sojourn zone

Subgranular zone

Rostral migratory stream (source area)

Caudate (head)

Reticular nucleus

Ventral Posterolateral nucleus

Internal capsule

Pulvinar

Lateral geniculate migration?

STF1 t1
STF1 t2
STF2
STF4 t2
STF4 t3?
STF5
STF6

Internal capsule (anterior limb)

THALAMUS

Lateral geniculate body

Occipital NEP and SVZ (thin)

Frontal (agranular) STF

Globus pallidus

Putamen

Basal nucleus of Meynert

Cerebral peduncle

Fornix

Orbitofrontal NEP and SVZ (thick)

Ventral striatum

Anterior commissure

Ansa lenticularis

Optic tract

Dentate migration

Ammonic migration and sojourn zone

Lateral migratory stream

Substantia innominata

Subgranular zone

VENTRAL HIPPOCAMPUS

CEREBELLUM (HEMISPHERE)

Primary olfactory cortex

Lateral olfactory tract

Medial nucleus

Central nucleus

Basal nucleus

Dentate gyrus

Hippocampal NEP

Simplex lobule (HVI)

External germinal layer

Cortical nucleus

Ammon's horn

Subiculum

Ansiform lobule (HVIIA)

Amygdaloid NEP

Entorhinal cortex

AMYGDALA

Lateral ventricle

Floccular peduncle

Germinal trigone

Middle and inferior cerebellar peduncles

Fourth ventricle (lateral recess with choroid plexus)

Germinal and transitional structures in *italics*

G/EP - glioepithelium/ependyma
GEP - glioepithelium
NEP - neuroepithelium
SVZ - subventricular zone

PLATE 30A
CR 145 mm, GW 17
Y37-63
Sagittal
Section 241

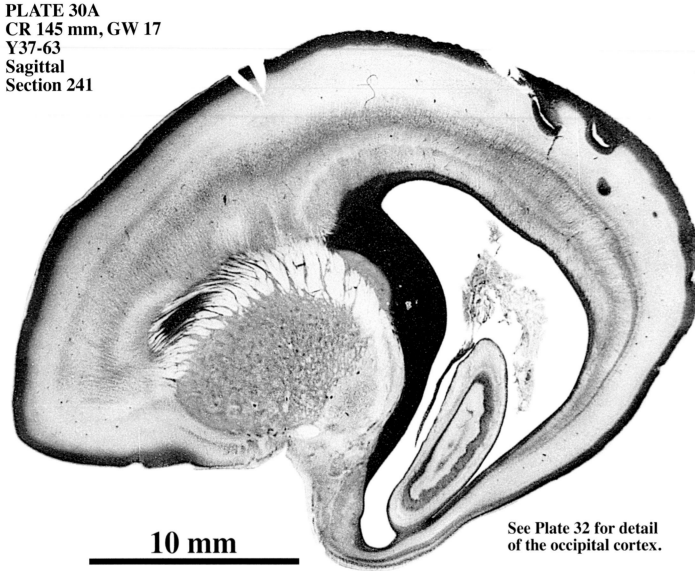

10 mm

See Plate 32 for detail of the occipital cortex.

LAYERS OF THE CORTICAL *STRATIFIED TRANSITIONAL FIELD (STF)*

STF1—Superficial fibrous layer with an early developmental stage *(t1)* when many cells are migrating through it, followed by a late stage *(t2)* with sparse cells. Endures as the subcortical white matter.

STF2—Upper cellular layer, the last sojourn zone before cells translocate to the cortical plate.

STF3—Honeycomb trilaminar matrix *(3a, 3b, 3c)* of cells and fibers found only in granular cortices.

STF4—Complex middle layer with three developmental stages:
t1– fibrous layer without interspersed cells;
t2– cells and fibers intermingle to form striations;
t3– fibers endure in the deep white matter.

STF5—Deep cellular layer, the first sojourn zone to appear outside the germinal matrix.

STF6—Late-forming deep layer of callosal fibers outside the germinal matrix.

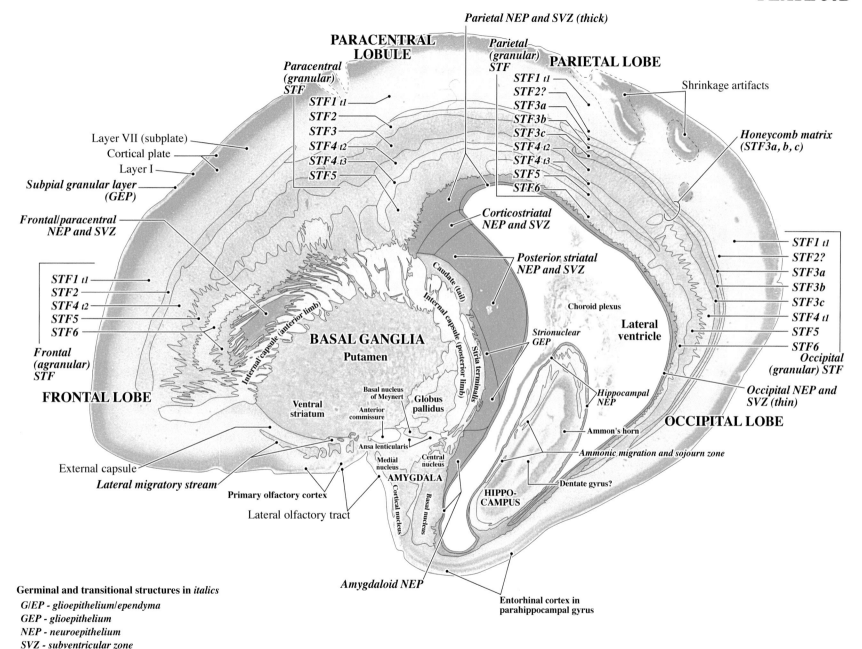

Parietal NEP and SVZ (thick)

PARACENTRAL LOBULE

Parietal (granular) STF

PARIETAL LOBE

Paracentral (granular) STF

STF1 t1
STF2
STF3
STF4 t2
STF4 t3
STF5

STF1 t1
STF2?
STF3a
STF3b
STF3c
STF4 t2
STF4 t3
STF5
STF6

Shrinkage artifacts

Layer VII (subplate)
Cortical plate
Layer I
Subpial granular layer (GEP)

Honeycomb matrix (STF3a, b, c)

Frontal/paracentral NEP and SVZ

Corticostriatal NEP and SVZ

Posterior striatal NEP and SVZ

STF1 t1
STF2
STF4 t2
STF5
STF6

STF1 t1
STF2?
STF3a
STF3b
STF3c
STF4 t1
STF5
STF6

Choroid plexus

Lateral ventricle

Occipital (granular) STF

Frontal (agranular) STF

BASAL GANGLIA
Putamen

Internal capsule (anterior limb)
Caudate (tail)
Internal capsule (posterior limb)
Stria terminalis

Strionuclear GEP

Occipital NEP and SVZ (thin)

FRONTAL LOBE

Ventral striatum

Basal nucleus of Meynert
Anterior commissure
Globus pallidus

Hippocampal NEP

Ammon's horn

OCCIPITAL LOBE

External capsule
Lateral migratory stream

Ansa lenticularis
Medial nucleus
Central nucleus
AMYGDALA

Ammonic migration and sojourn zone

Primary olfactory cortex
Lateral olfactory tract

Cortical nucleus
Basal nucleus

Dentate gyrus?

HIPPO-CAMPUS

Amygdaloid NEP

Entorhinal cortex in parahippocampal gyrus

Germinal and transitional structures in italics
G/EP - glioepithelium/ependyma
GEP - glioepithelium
NEP - neuroepithelium
SVZ - subventricular zone

**PLATE 31
CR 145 mm
GW 17, Y37-63
Sagittal
Sections 140/141
PARACENTRAL
CORTEX**

A. Section 140

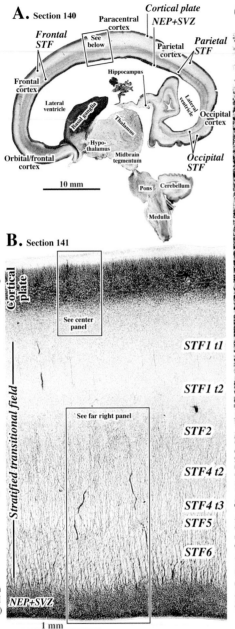

*Frontal
STF*

Paracentral
cortex

*Cortical plate
NEP+SVZ*

Frontal
cortex

Parietal
cortex

*Parietal
STF*

Hippocampus

Lateral
ventricle

Lateral
ventricle

Occipital
cortex

Basal ganglia

Thalamus

Orbital/frontal
cortex

Hypo-
thalamus

Midbrain
tegmentum

*Occipital
STF*

Pons   Cerebellum

Medulla

10 mm

A. A low-magnification view
of Section 140. The entire
section is labeled in **Plate 27.**
The box indicates the area of
the paracentral cortex from
Section 141 enlarged in **B.**

B. The *stratified transitional
field* (*STF*) is thick compared to
the cortical plate, the
neuroepithelium (*NEP*) and the
subventricular zone (*SVZ*) in
this slice of the paracentral
cortex from the box in **A.**
Actual thicknesses are:
NEP+SVZ=0.316 mm,
Cortical plate=0.353 mm,
STF=2.732 mm
The *STF* is 8.65 times thicker
than the *NEP+SVZ* and 7.74
times thicker than the cortical
plate.
Note that the upper part of
STF1 is in the *t1* stage with
higher cell density. The upper
rectangle shows the area of the
cortical plate enlarged in **C.**
The lower rectangle shows the
area of the *NEP+SVZ* and deep
STF enlarged in **D.**

C. The paracentral cortical
plate and subplate from the
upper box in **B. Roman
numerals** indicate the cortical
layers (**I-VII**). Cells are
densely packed throughout the
cortical plate, and no distinction
can be made between Layers VI
through II. Layer VII (subplate)
is much less dense.

B. Section 141

Cortical
plate

See center
panel

See far right panel

Stratified transitional field

STF1 t1

STF1 t2

STF2

STF4 t2

STF4 t3
STF5

STF6

NEP+SVZ

1 mm

C. Section 141

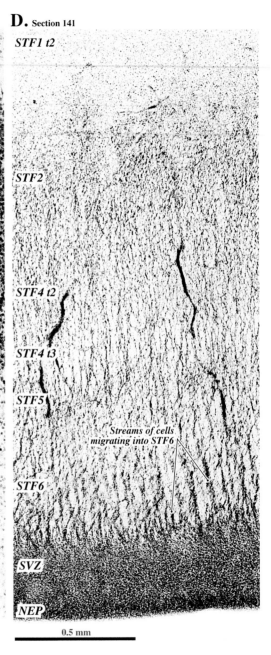

Subpial granular zone
(a transient glial matrix)

I

UNDEFINED LAYERS VI THROUGH II

VII

STF1 t1

0.1 mm

D. Section 141

STF1 t2

STF2

STF4 t2

STF4 t3

STF5

STF6

*Streams of cells
migrating into STF6*

SVZ

NEP

0.5 mm

D. The paracentral *STF, NEP,* and *SVZ* from the lower box in **B.** *STF2* and *STF5* have
relatively high cell density. *STF4* can be subdivided into a deep mature stage (*t3*) when fibers
predominate and a superficial less mature stage when cells and fibers intermingle (*t2*).

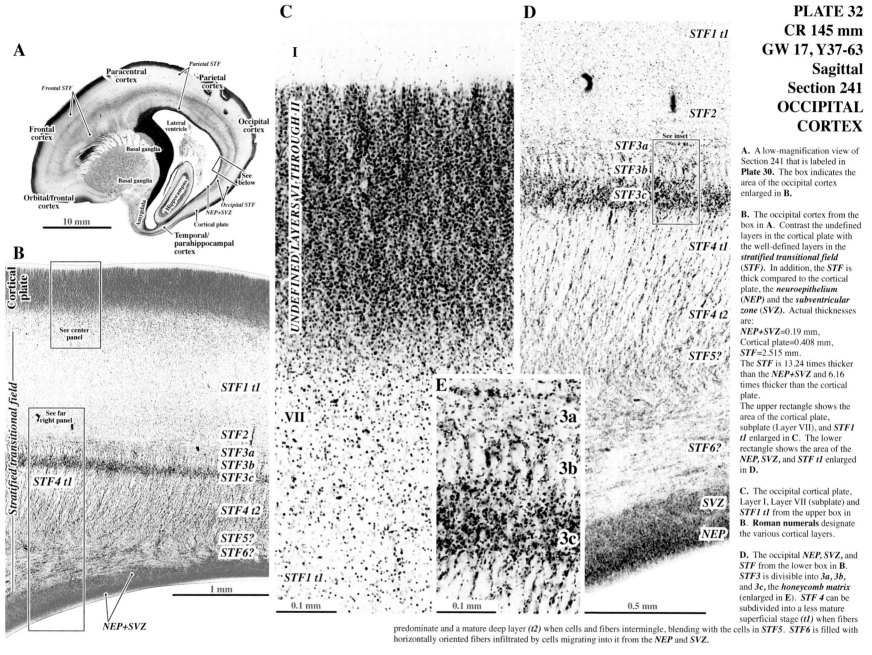

A

Frontal STF
Paracentral cortex
Parietal STF
Parietal cortex
Frontal cortex
Lateral ventricle
Occipital cortex
Orbital/frontal cortex
Basal ganglia
Basal ganglia
Hippocampus
See below
Occipital STF
NEP+SVZ
Cortical plate
Amygdala
Temporal/ parahippocampal cortex

10 mm

B

Cortical plate
See center panel
STF1 t1
See far right panel
Stratified transitional field
STF2
STF3a
STF3b
STF3c
STF4 t1
STF4 t2
STF5?
STF6?
NEP+SVZ

1 mm

C

I
UNDEFINED LAYERS VI THROUGH II
VII
STF1 t1

0.1 mm

E

3a
3b
3c

0.1 mm

D

STF1 t1
STF2
STF3a
STF3b
STF3c
See inset
STF4 t1
STF4 t2
STF5?
STF6?
SVZ
NEP

0.5 mm

PLATE 32
CR 145 mm
GW 17, Y37-63
Sagittal
Section 241
OCCIPITAL
CORTEX

A. A low-magnification view of Section 241 that is labeled in **Plate 30.** The box indicates the area of the occipital cortex enlarged in **B.**

B. The occipital cortex from the box in **A.** Contrast the undefined layers in the cortical plate with the well-defined layers in the *stratified transitional field* (*STF*). In addition, the *STF* is thick compared to the cortical plate, the *neuroepithelium* (*NEP*) and the *subventricular zone* (*SVZ*). Actual thicknesses are:
NEP+SVZ=0.19 mm,
Cortical plate=0.408 mm,
STF=2.515 mm.
The *STF* is 13.24 times thicker than the *NEP+SVZ* and 6.16 times thicker than the cortical plate.
The upper rectangle shows the area of the cortical plate, subplate (Layer VII), and *STF1 t1* enlarged in **C.** The lower rectangle shows the area of the *NEP, SVZ,* and *STF t1* enlarged in **D.**

C. The occipital cortical plate, Layer I, Layer VII (subplate) and *STF1 t1* from the upper box in **B. Roman numerals** designate the various cortical layers.

D. The occipital *NEP, SVZ,* and *STF* from the lower box in **B.** *STF3* is divisible into *3a, 3b,* and *3c,* the *honeycomb matrix* (enlarged in **E**). *STF 4* can be subdivided into a less mature superficial stage *(t1)* when fibers predominate and a mature deep layer *(t2)* when cells and fibers intermingle, blending with the cells in *STF5*. *STF6* is filled with horizontally oriented fibers infiltrated by cells migrating into it from the *NEP* and *SVZ*.

E. The honeycomb matrix is subdivided into diffusely scattered cells in *STF3a*, vertically oriented columns of cells in between fibers in *STF3b*, and high cell density in *STF3c*.

70

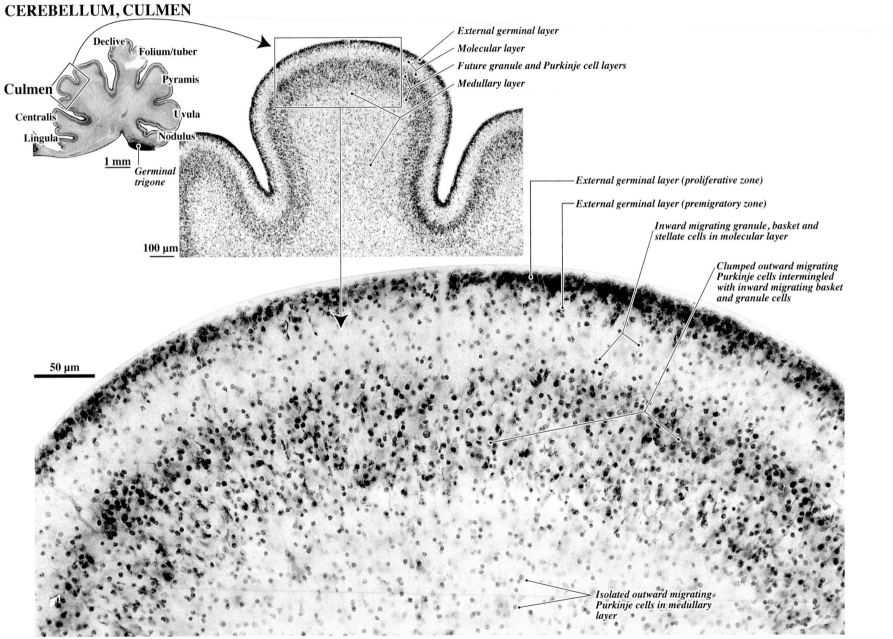

Declive
Folium/tuber
Pyramis
Culmen
Centralis
Uvula
Lingula
Nodulus
1 mm
Germinal
trigone

External germinal layer
Molecular layer
Future granule and Purkinje cell layers
Medullary layer

100 µm

External germinal layer (proliferative zone)

External germinal layer (premigratory zone)

Inward migrating granule, basket and
stellate cells in molecular layer

Clumped outward migrating
Purkinje cells intermingled
with inward migrating basket
and granule cells

50 µm

Isolated outward migrating
Purkinje cells in medullary
layer

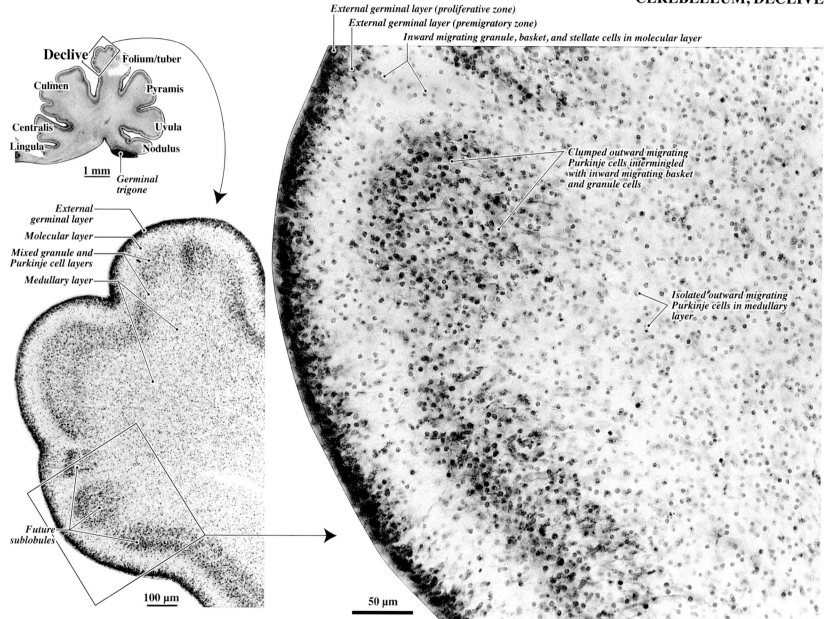

Declive

Folium/tuber

Culmen

Pyramis

Centralis

Uvula

Lingula

Nodulus

1 mm

Germinal trigone

External germinal layer

Molecular layer

Mixed granule and Purkinje cell layers

Medullary layer

Future sublobules

100 µm

External germinal layer (proliferative zone)
External germinal layer (premigratory zone)
Inward migrating granule, basket, and stellate cells in molecular layer

Clumped outward migrating Purkinje cells intermingled with inward migrating basket and granule cells

Isolated outward migrating Purkinje cells in medullary layer

50 µm

PART IV: Y15-63
CR 150 mm (GW 17)
Frontal

This specimen is a stillborn fetus of undetermined sex (Yakovlev case number BR-15-63, referred to as Y15-63). The age was originally estimated to be at gestational week (GW) 19, but the crown-rump length (CR 150 mm) and the general level of brain maturation place it more probably at GW17. The brain was cut in the frontal plane in 979 35-μm thick sections and is classified as a Normative Control in the Yakovlev Collection (Haleem, 1990). Since there is no photograph of this brain before it was embedded and cut, we turned to the comprehensive atlas that Retzius published in 1896 showing whole fetal brains in medial, lateral, superior, and inferior views and midline sagittally cut brains. **Figure 9**, taken from Retzius (1896), shows the exterior of a brain from a specimen that is at the same age as Y15-63, along with the approximate cutting angle of the sections. Photographs of 15 Nissl-stained entire sections are illustrated in **Plates 35–49**. **Plates 50–62** show various high-magnification views of the brain core. Different areas of the cerebral cortex are shown in greater detail in **Plates 63–66**.

The *cortical neuroepithelium and subventricular zone* is generating neocortical neurons mainly for superficial layers. The *stratified transitional fields* in all lobes of the cerebral cortex are filled with migrating and sojourning neurons and show distinct regional heterogeneity between granular (future sensory) and agranular (future motor) areas. The cerebral cortex is smooth except for the lateral fissure and the calcarine sulcus; the narrow invaginations seen in anterior and posterior regions are shrinkage artifacts produced during fixation. Many neurons, glia, and their mitotic precursor cells are still migrating through the olfactory peduncle toward the olfactory bulb (*rostral migratory stream*). In anterolateral parts of the cerebral cortex, "rivulets" of neurons and glia are numerous in the *lateral migratory stream* heading toward the insular cortex, primary olfactory cortex, temporal cortex, and basolateral parts of the amygdaloid complex. In the hippocampus, the *ammonic migration*, *dentate migration*, and *subgranular zone* are prominent, but it still has a thin *neuroepithelium* at the ventricle where some neurons are still being generated. Part of the hippocampus is positioned dorsal to the thalamus because the neocortical parts of the cerebral cortex have not grown enough to push it completely into the temporal lobe. A massive *neuroepithelium/subventricular zone* overlies the nucleus accumbens and striatum where neurons (and glia) are being generated. The *striatal neuroepithelium and subventricular zone* has indistinct subdivisions, but the *strionuclear glioepithelium* forms a definite bulge at its medial edge. That glioepithelium may provide oligodendrocytes for the stria terminalis, internal capsule, stria medullaris, and contribute oligodendroglial precursors to the *perifascicular glioepithelium* bordering the optic tract and the cerebral peduncle. The septum has a *glioepithelium/ependyma* at the ventricular surface and its nuclei are generally well-defined.

Neurons in most diencephalic structures appear to be settled; the major exceptions are the immature appearance of the lateral and medial geniculate bodies in the posterior thalamus and the hypothalamic medial mammillary body. The third ventricle is lined by a thin *glioepithelium/ependyma* that shows small invaginations and evaginations that mark previous sites where fate-restricted neuroepithelial patches generated neurons for specific nuclei. In the midbrain, pons, and medulla, there is a convoluted *glioepithelium/ependyma* lining the cerebral aqueduct and fourth ventricle that probably marks sites of past fate-restricted neuroepithelia. In the medulla, pontine gray neurons are still migrating in the *anterior extramural migratory stream* toward the basal pons.

The cerebellum is enlarging, especially the hemispheres. Major lobules are identifiable in the midline vermis. The deep nuclei are in place beneath the cortex. The entire surface of the cerebellar cortex is covered by the prominent *external germinal layer (egl)* that is actively producing basket, stellate, and granule cells. Lamination in the cortex is less definite, except for a thin molecular layer beneath the *egl*. Nearly all Purkinje cells are migrating. The *germinal trigone* is large at the base of the nodulus and along the floccular peduncle.

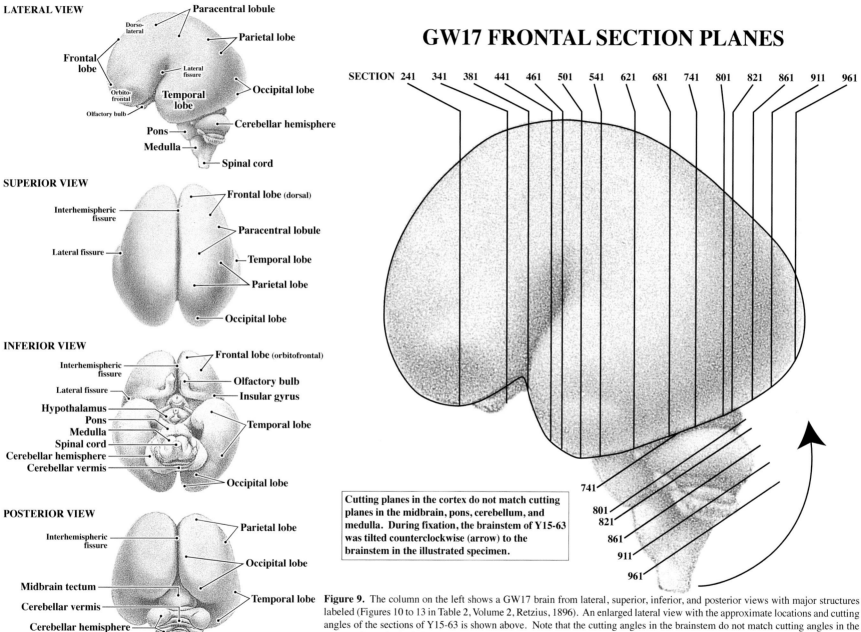

LATERAL VIEW

Paracentral lobule
Dorso-lateral
Parietal lobe
Frontal lobe
Lateral fissure
Occipital lobe
Orbito-frontal
Temporal lobe
Olfactory bulb
Cerebellar hemisphere
Pons
Medulla
Spinal cord

SUPERIOR VIEW

Frontal lobe (dorsal)
Interhemispheric fissure
Paracentral lobule
Lateral fissure
Temporal lobe
Parietal lobe
Occipital lobe

INFERIOR VIEW

Frontal lobe (orbitofrontal)
Interhemispheric fissure
Lateral fissure
Olfactory bulb
Insular gyrus
Hypothalamus
Pons
Medulla
Spinal cord
Temporal lobe
Cerebellar hemisphere
Cerebellar vermis
Occipital lobe

POSTERIOR VIEW

Parietal lobe
Interhemispheric fissure
Occipital lobe
Midbrain tectum
Cerebellar vermis
Temporal lobe
Cerebellar hemisphere
Medulla
Spinal cord

GW17 FRONTAL SECTION PLANES

SECTION 241 341 381 441 461 501 541 621 681 741 801 821 861 911 961

741
801
821
861
911
961

Cutting planes in the cortex do not match cutting planes in the midbrain, pons, cerebellum, and medulla. During fixation, the brainstem of Y15-63 was tilted counterclockwise (arrow) to the brainstem in the illustrated specimen.

Figure 9. The column on the left shows a GW17 brain from lateral, superior, inferior, and posterior views with major structures labeled (Figures 10 to 13 in Table 2, Volume 2, Retzius, 1896). An enlarged lateral view with the approximate locations and cutting angles of the sections of Y15-63 is shown above. Note that the cutting angles in the brainstem do not match cutting angles in the cortex. Once the brain has been dissected from the skull, the brainstem has no support to maintain a constant angle of downward extension from the cortex. Consequently, the brainstem can easily be flexed either forward or backward during fixation. In Y15-63, the brainstem was flexed backward and upward (arrow) to tuck under the cerebral hemispheres.

74

PLATE 35A
CR 150 mm, GW 17, Y15-63
Frontal
Section 241

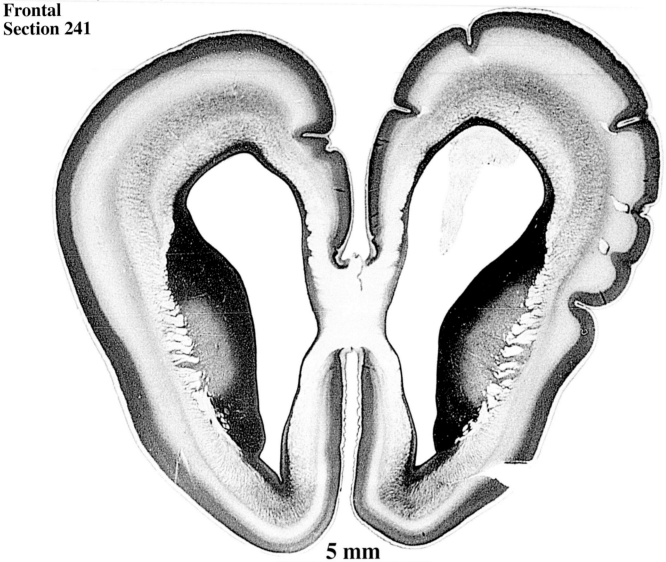

5 mm

LAYERS OF THE CORTICAL
STRATIFIED TRANSITIONAL
FIELD (STF)

STF1—Superficial fibrous
layer with an early
developmental stage *(t1)*
when many cells are
migrating through it, followed
by a late stage *(t2)* with sparse
cells. Endures as the
subcortical white matter.

STF2—Upper cellular layer,
the last sojourn zone before
cells translocate to the cortical
plate.

STF4—Complex middle
layer with three
developmental stages:
t1– fibrous layer without
interspersed cells;
t2– cells and fibers
intermingle to form striations;
t3– fibers endure in the deep
white matter.

STF5—Deep cellular layer,
the first sojourn zone to
appear outside the germinal
matrix.

STF6—Late-forming deep
layer of callosal fibers outside
the germinal matrix.

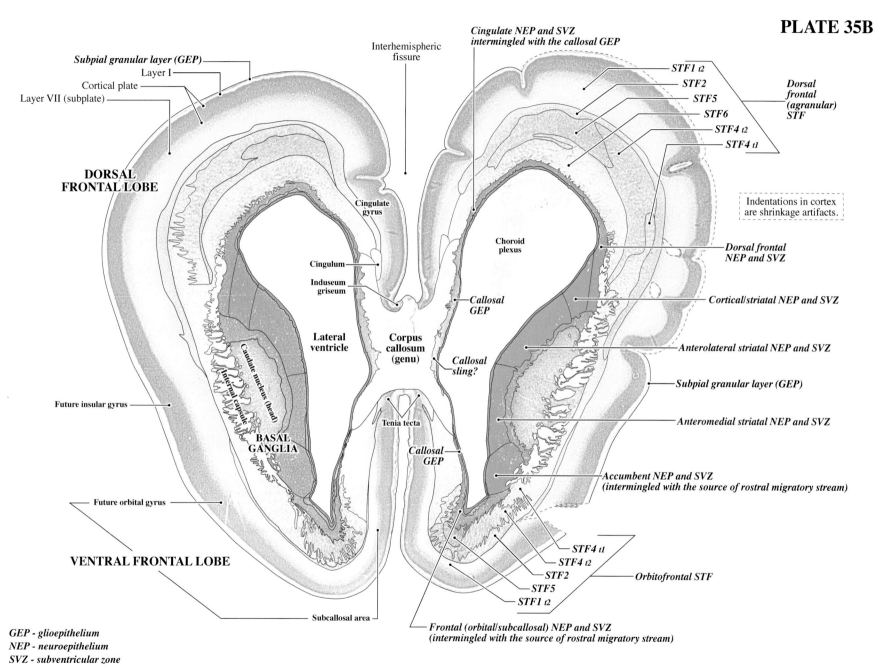

Cingulate NEP and SVZ
intermingled with the callosal GEP

STF1 t2

STF2

STF5

STF6

STF4 t2

STF4 t1

*Dorsal
frontal
(agranular)
STF*

Subpial granular layer (GEP)

Layer I

Cortical plate

Layer VII (subplate)

Interhemispheric
fissure

Indentations in cortex
are shrinkage artifacts.

**DORSAL
FRONTAL LOBE**

Cingulate
gyrus

Choroid
plexus

*Dorsal frontal
NEP and SVZ*

Cingulum

Induseum
griseum

*Callosal
GEP*

Cortical/striatal NEP and SVZ

**Lateral
ventricle**

**Corpus
callosum
(genu)**

*Callosal
sling?*

Anterolateral striatal NEP and SVZ

Caudate nucleus (head)

Internal capsule

Subpial granular layer (GEP)

Future insular gyrus

Anteromedial striatal NEP and SVZ

**BASAL
GANGLIA**

Tenia tecta

*Callosal
GEP*

*Accumbent NEP and SVZ
(intermingled with the source of rostral migratory stream)*

Future orbital gyrus

STF4 t1

STF4 t2

STF2

STF5

STF1 t2

Orbitofrontal STF

VENTRAL FRONTAL LOBE

Subcallosal area

*Frontal (orbital/subcallosal) NEP and SVZ
(intermingled with the source of rostral migratory stream)*

GEP - *glioepithelium*

NEP - *neuroepithelium*

SVZ - *subventricular zone*

Germinal and transitional structures in *italics*

PLATE 36A
CR 150 mm, GW 17, Y15-63
Frontal
Section 341

See this area of cortex
in Plate 63.

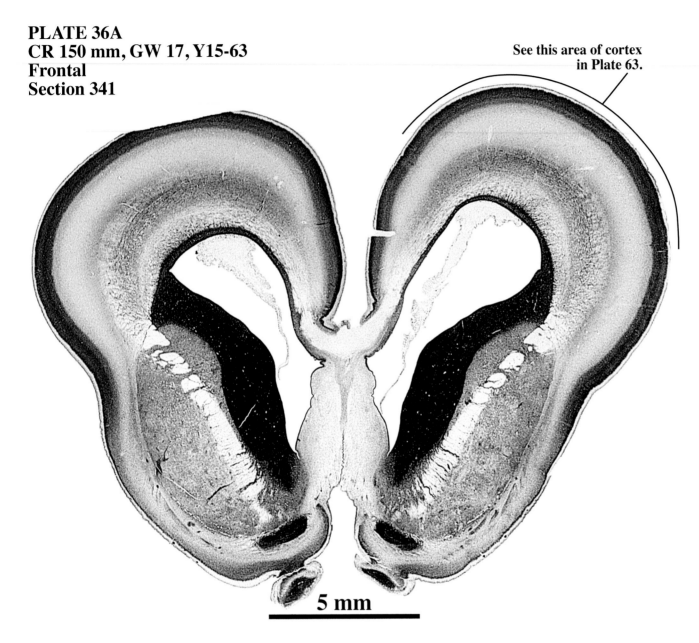

5 mm

LAYERS OF THE CORTICAL
STRATIFIED TRANSITIONAL
FIELD (STF)

STF1—Superficial fibrous
layer with an early
developmental stage *(t1)*
when many cells are
migrating through it, followed
by a late stage *(t2)* with sparse
cells. Endures as the
subcortical white matter.

STF2—Upper cellular layer,
the last sojourn zone before
cells translocate to the cortical
plate.

STF4—Complex middle
layer with three
developmental stages:
t1– fibrous layer without
interspersed cells;
t2– cells and fibers
intermingle to form striations;
t3– fibers endure in the deep
white matter.

STF5—Deep cellular layer,
the first sojourn zone to
appear outside the germinal
matrix.

STF6—Late-forming deep
layer of callosal fibers outside
the germinal matrix.

See detail of the brain core in Plates 50A and B.

77

PLATE 36B

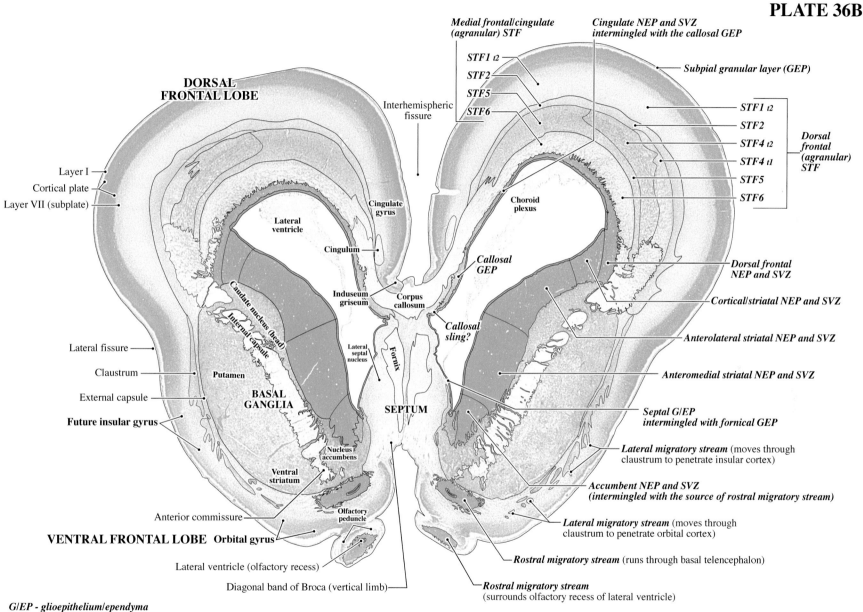

G/EP - glioepithelium/ependyma
GEP - glioepithelium
NEP - neuroepithelium
SVZ - subventricular zone

Germinal and transitional structures in *italics*

PLATE 37A
CR 150 mm, GW 17, Y15-63
Frontal
Section 381

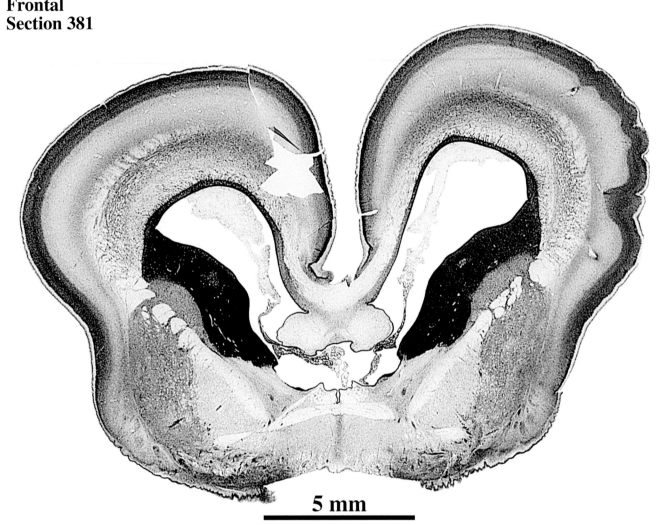

5 mm

LAYERS OF THE CORTICAL
STRATIFIED TRANSITIONAL
FIELD (STF)

STF1—Superficial fibrous
layer with an early
developmental stage *(t1)*
when many cells are
migrating through it, followed
by a late stage *(t2)* with sparse
cells. Endures as the
subcortical white matter.

STF2—Upper cellular layer,
the last sojourn zone before
cells translocate to the cortical
plate.

STF4—Complex middle
layer with three
developmental stages:
t1– fibrous layer without
interspersed cells;
t2– cells and fibers
intermingle to form striations;
t3– fibers endure in the deep
white matter.

STF5—Deep cellular layer,
the first sojourn zone to
appear outside the germinal
matrix.

STF6—Late-forming deep
layer of callosal fibers outside
the germinal matrix.

See detail of brain core in Plates 51A and B.

PLATE 37B

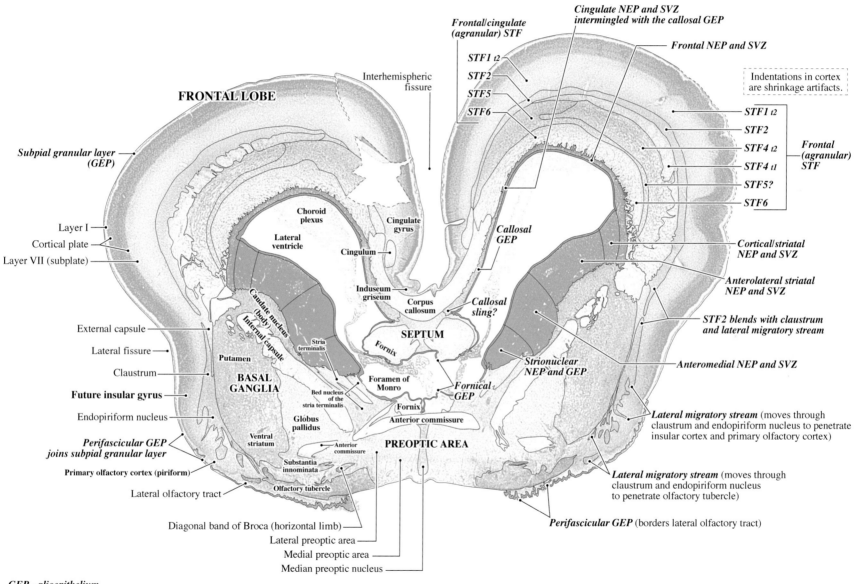

Cingulate NEP and SVZ
intermingled with the callosal GEP

*Frontal/cingulate
(agranular) STF*

Frontal NEP and SVZ

STF1 t2

STF2

STF5

STF6

Indentations in cortex
are shrinkage artifacts.

STF1 t2

STF2

STF4 t2

STF4 t1

STF5?

STF6

*Frontal
(agranular)
STF*

FRONTAL LOBE

Interhemispheric
fissure

*Subpial granular layer
(GEP)*

Choroid
plexus

Cingulate
gyrus

*Callosal
GEP*

Layer I

Lateral
ventricle

Cingulum

*Cortical/striatal
NEP and SVZ*

Cortical plate

Layer VII (subplate)

Induseum
griseum

Corpus
callosum

*Callosal
sling?*

*Anterolateral striatal
NEP and SVZ*

Caudate
nucleus
(body)

SEPTUM

*STF2 blends with claustrum
and lateral migratory stream*

External capsule

Internal
capsule

Stria
terminalis

Fornix

Lateral fissure

Claustrum

Foramen of
Monro

*Fornical
GEP*

*Strionuclear
NEP and GEP*

Anteromedial NEP and SVZ

Putamen

**BASAL
GANGLIA**

Future insular gyrus

Bed nucleus
of the
stria terminalis

Fornix

Endopiriform nucleus

Globus
pallidus

Anterior commissure

Lateral migratory stream (moves through
claustrum and endopiriform nucleus to penetrate
insular cortex and primary olfactory cortex)

*Perifascicular GEP
joins subpial granular layer*

Ventral
striatum

Anterior
commissure

PREOPTIC AREA

Primary olfactory cortex (piriform)

Substantia
innominata

Lateral migratory stream (moves through
claustrum and endopiriform nucleus
to penetrate olfactory tubercle)

Lateral olfactory tract

Olfactory tubercle

Diagonal band of Broca (horizontal limb)

Perifascicular GEP (borders lateral olfactory tract)

Lateral preoptic area

Medial preoptic area

Median preoptic nucleus

GEP - glioepithelium
NEP - neuroepithelium
SVZ - subventricular zone

Germinal and transitional structures in *italics*

**PLATE 38A
CR 150 mm, GW 17, Y15-63
Frontal
Section 441**

See this area of cortex from
Section 451 in Plate 64.

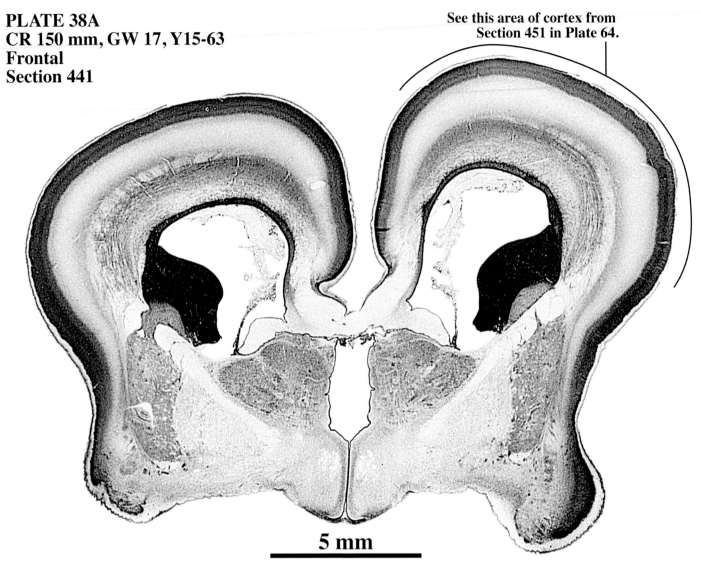

LAYERS OF THE CORTICAL
STRATIFIED TRANSITIONAL
FIELD (STF)

STF1—Superficial fibrous
layer with an early
developmental stage *(t1)*
when many cells are
migrating through it, followed
by a late stage *(t2)* with sparse
cells. Endures as the
subcortical white matter.

STF2—Upper cellular layer,
the last sojourn zone before
cells translocate to the cortical
plate.

STF4—Complex middle
layer with three
developmental stages:
t1– fibrous layer without
interspersed cells;
t2– cells and fibers
intermingle to form striations;
t3– fibers endure in the deep
white matter.

STF5—Deep cellular layer,
the first sojourn zone to
appear outside the germinal
matrix.

STF6—Late-forming deep
layer of callosal fibers outside
the germinal matrix.

5 mm

See detail of brain core in Plates 52A and B.

PLATE 38B

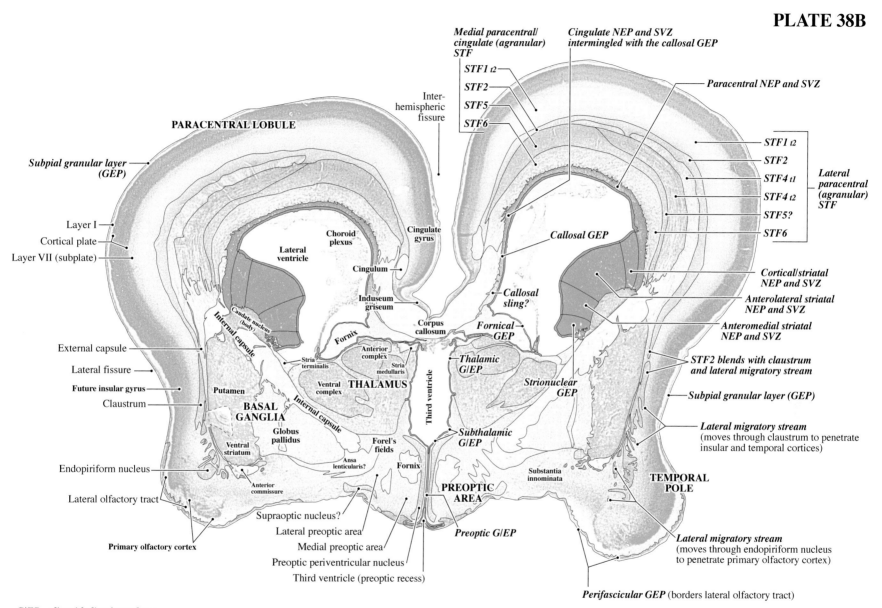

*Medial paracentral/
cingulate (agranular)
STF*

STF1 t2
STF2
STF5
STF6

*Cingulate NEP and SVZ
intermingled with the callosal GEP*

Paracentral NEP and SVZ

Inter-
hemispheric
fissure

PARACENTRAL LOBULE

STF1 t2
STF2
STF4 t1
STF4 t2
STF5?
STF6

*Lateral
paracentral
(agranular)
STF*

*Subpial granular layer
(GEP)*

Layer I
Cortical plate
Layer VII (subplate)

Choroid
plexus

Cingulate
gyrus

Callosal GEP

Lateral
ventricle

Cingulum

*Cortical/striatal
NEP and SVZ*

Induseum
griseum

*Anterolateral striatal
NEP and SVZ*

Caudate nucleus
(body)

Corpus
callosum

*Callosal
sling?*

*Anteromedial striatal
NEP and SVZ*

Internal capsule

Fornix

*Fornical
GEP*

External capsule

Stria
terminalis

Anterior
complex

Stria
medullaris

*Thalamic
G/EP*

*STF2 blends with claustrum
and lateral migratory stream*

Lateral fissure

Ventral
complex

THALAMUS

*Strionuclear
GEP*

Subpial granular layer (GEP)

Future insular gyrus

Claustrum

Putamen

**BASAL
GANGLIA**

Internal capsule

Third ventricle

Lateral migratory stream
(moves through claustrum to penetrate
insular and temporal cortices)

Globus
pallidus

Forel's
fields

*Subthalamic
G/EP*

Ventral
striatum

Endopiriform nucleus

Ansa
lenticularis?

Fornix

Substantia
innominata

**TEMPORAL
POLE**

Anterior
commissure

**PREOPTIC
AREA**

Lateral olfactory tract

Supraoptic nucleus?

Lateral preoptic area

Medial preoptic area

Preoptic G/EP

Lateral migratory stream
(moves through endopiriform nucleus
to penetrate primary olfactory cortex)

Primary olfactory cortex

Preoptic periventricular nucleus

Third ventricle (preoptic recess)

Perifascicular GEP (borders lateral olfactory tract)

G/EP - glioepithelium/ependyma
GEP - glioepithelium
NEP - neuroepithelium
SVZ - subventricular zone

Germinal and transitional structures in *italics*

PLATE 39A
CR 150 mm, GW 17, Y15-63
Frontal
Section 461

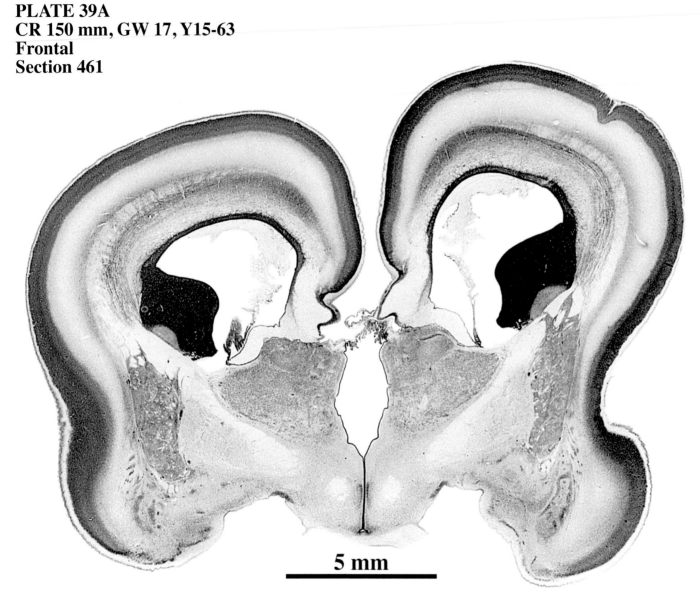

5 mm

LAYERS OF THE CORTICAL
STRATIFIED TRANSITIONAL
FIELD (STF)

STF1—Superficial fibrous layer with an early developmental stage *(t1)* when many cells are migrating through it, followed by a late stage *(t2)* with sparse cells. Endures as the subcortical white matter.

STF2—Upper cellular layer, the last sojourn zone before cells translocate to the cortical plate.

STF3—Honeycomb trilaminar matrix *(3a, 3b, 3c)* of cells and fibers found only in granular cortices.

STF4—Complex middle layer with three developmental stages:
t1– fibrous layer without interspersed cells;
t2– cells and fibers intermingle to form striations;
t3– fibers endure in the deep white matter.

STF5—Deep cellular layer, the first sojourn zone to appear outside the germinal matrix.

STF6—Late-forming deep layer of callosal fibers outside the germinal matrix.

See detail of brain core in Plates 53A and B.

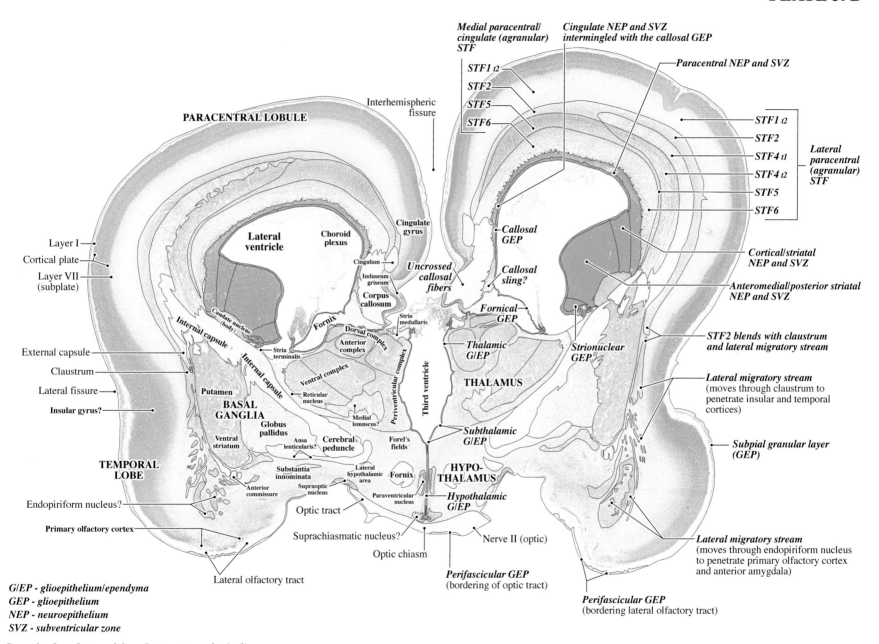

Medial paracentral/cingulate (agranular) STF

Cingulate NEP and SVZ intermingled with the callosal GEP

Paracentral NEP and SVZ

STF1 t2
STF2
STF5
STF6

Interhemispheric fissure

PARACENTRAL LOBULE

STF1 t2
STF2
STF4 t1
STF4 t2
STF5
STF6

Lateral paracentral (agranular) STF

Lateral ventricle

Choroid plexus

Cingulate gyrus

Callosal GEP

Cortical/striatal NEP and SVZ

Layer I
Cortical plate
Layer VII (subplate)

Cingulum

Induseum griseum

Corpus callosum

Uncrossed callosal fibers

Callosal sling?

Anteromedial/posterior striatal NEP and SVZ

Caudate nucleus (body)

Fornix

Stria medullaris

Fornical GEP

Internal capsule

Dorsal complex

Anterior complex

Stria terminalis

Thalamic G/EP

Strionuclear GEP

STF2 blends with claustrum and lateral migratory stream

External capsule
Claustrum
Lateral fissure
Insular gyrus?

Internal capsule

Ventral complex

Periventricular complex

Third ventricle

THALAMUS

Lateral migratory stream (moves through claustrum to penetrate insular and temporal cortices)

Putamen
BASAL GANGLIA
Globus pallidus

Reticular nucleus

Medial lemniscus?

Ventral striatum

Ansa lenticularis?

Cerebral peduncle

Forel's fields

Subthalamic G/EP

Subpial granular layer (GEP)

TEMPORAL LOBE

Substantia innominata

Lateral hypothalamic area

Fornix

HYPO-THALAMUS

Anterior commissure

Supraoptic nucleus

Paraventricular nucleus

Hypothalamic G/EP

Endopiriform nucleus?

Optic tract

Suprachiasmatic nucleus?

Optic chiasm

Nerve II (optic)

Lateral migratory stream (moves through endopiriform nucleus to penetrate primary olfactory cortex and anterior amygdala)

Primary olfactory cortex

Lateral olfactory tract

Perifascicular GEP (bordering of optic tract)

Perifascicular GEP (bordering lateral olfactory tract)

G/EP - glioepithelium/ependyma
GEP - glioepithelium
NEP - neuroepithelium
SVZ - subventricular zone

Germinal and transitional structures in *italics*

PLATE 40A
CR 150 mm, GW 17, Y15-63
Frontal
Section 501

See this area of cortex —
in Plate 65.

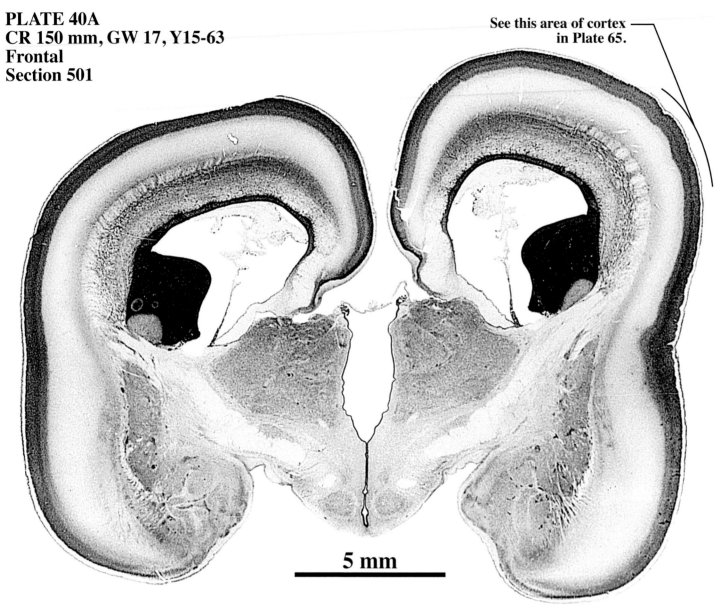

5 mm

See detail of brain core in Plates 54A and B.

LAYERS OF THE CORTICAL
STRATIFIED TRANSITIONAL
FIELD (STF)

STF1—Superficial fibrous layer with an early developmental stage *(t1)* when many cells are migrating through it, followed by a late stage *(t2)* with sparse cells. Endures as the subcortical white matter.

STF2—Upper cellular layer, the last sojourn zone before cells translocate to the cortical plate.

STF3—Honeycomb trilaminar matrix *(3a, 3b, 3c)* of cells and fibers found only in granular cortices.

STF4—Complex middle layer with three developmental stages:
t1– fibrous layer without interspersed cells;
t2– cells and fibers intermingle to form striations;
t3– fibers endure in the deep white matter.

STF5—Deep cellular layer, the first sojourn zone to appear outside the germinal matrix.

STF6—Late-forming deep layer of callosal fibers outside the germinal matrix.

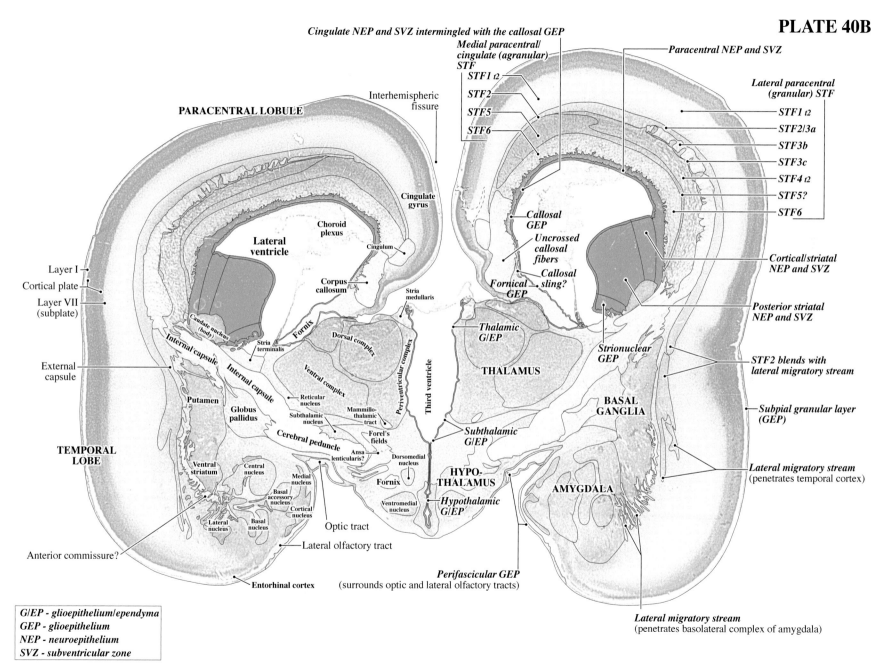

Cingulate NEP and SVZ intermingled with the callosal GEP

Medial paracentral/
cingulate (agranular)
STF
Paracentral NEP and SVZ

STF1 t2
STF2
STF5
STF6

Lateral paracentral
(granular) STF

STF1 t2
STF2/3a
STF3b
STF3c
STF4 t2
STF5?
STF6

PARACENTRAL LOBULE

Interhemispheric
fissure

Cingulate
gyrus

Choroid
plexus

Lateral
ventricle

Cingulum

Corpus
callosum

Callosal
GEP

Uncrossed
callosal
fibers

Callosal
sling?

Cortical/striatal
NEP and SVZ

Stria
medullaris

Fornical
GEP

Layer I
Cortical plate
Layer VII
(subplate)

Posterior striatal
NEP and SVZ

Caudate nucleus
(body)

Stria
terminalis

Fornix

Dorsal complex

Periventricular complex

Thalamic
G/EP

THALAMUS

Strionuclear
GEP

STF2 blends with
lateral migratory stream

External
capsule

Internal capsule

Internal capsule

Ventral complex

Third ventricle

Reticular
nucleus

Putamen

Globus
pallidus

Subthalamic
nucleus

Mammillo-
thalamic
tract

Forel's
fields

Subthalamic
G/EP

BASAL
GANGLIA

Subpial granular layer
(GEP)

TEMPORAL
LOBE

Cerebral peduncle

Ansa
lenticularis?

Dorsomedial
nucleus

HYPO-
THALAMUS

Lateral migratory stream
(penetrates temporal cortex)

Ventral
striatum

Central
nucleus

Medial
nucleus

Fornix

Ventromedial
nucleus

Hypothalamic
G/EP

AMYGDALA

Lateral
nucleus

Basal
accessory
nucleus

Basal
nucleus

Cortical
nucleus

Optic tract

Anterior commissure?

Lateral olfactory tract

Perifascicular GEP
(surrounds optic and lateral olfactory tracts)

Entorhinal cortex

Lateral migratory stream
(penetrates basolateral complex of amygdala)

G/EP - glioepithelium/ependyma
GEP - glioepithelium
NEP - neuroepithelium
SVZ - subventricular zone

Germinal and transitional structures in *italics*

PLATE 41A
CR 150 mm, GW 17, Y15-63
Frontal
Section 541

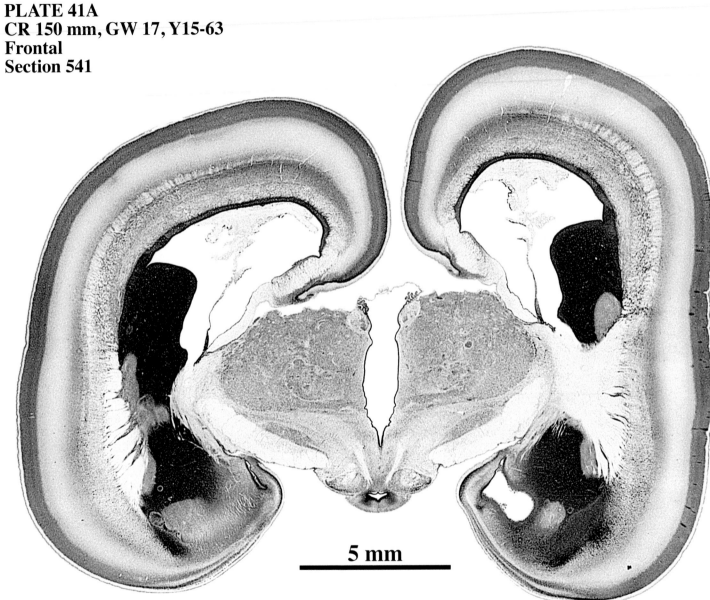

5 mm

LAYERS OF THE CORTICAL *STRATIFIED TRANSITIONAL FIELD (STF)*

STF1—Superficial fibrous layer with an early developmental stage *(t1)* when many cells are migrating through it, followed by a late stage *(t2)* with sparse cells. Endures as the subcortical white matter.

STF2—Upper cellular layer, the last sojourn zone before cells translocate to the cortical plate.

STF3—Honeycomb trilaminar matrix *(3a, 3b, 3c)* of cells and fibers found only in granular cortices.

STF4—Complex middle layer with three developmental stages: *t1*– fibrous layer without interspersed cells; *t2*– cells and fibers intermingle to form striations; *t3*– fibers endure in the deep white matter.

STF5—Deep cellular layer, the first sojourn zone to appear outside the germinal matrix.

STF6—Late-forming deep layer of callosal fibers outside the germinal matrix.

See detail of brain core in Plates 55A and B.

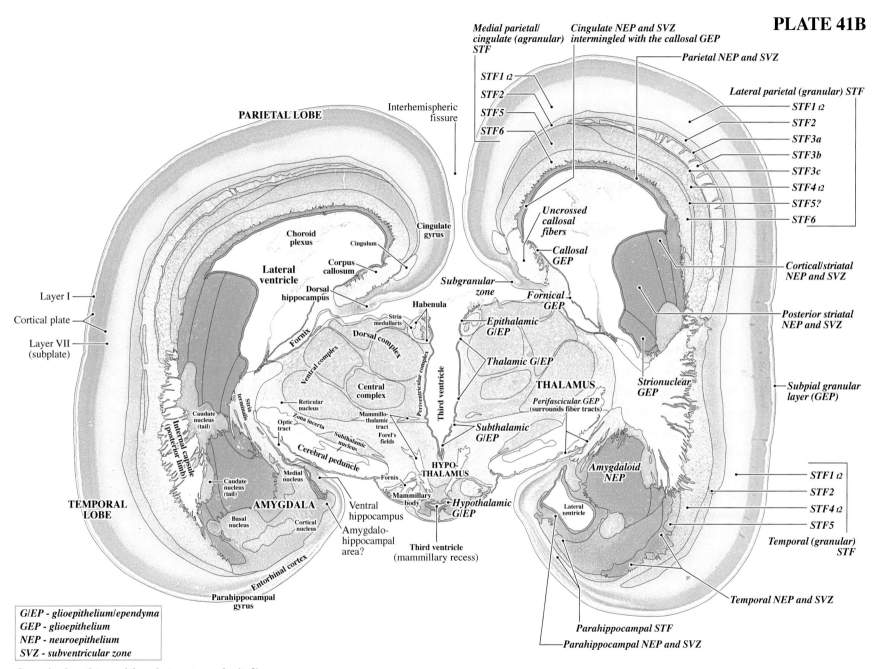

PLATE 41B

Medial parietal/
cingulate (agranular)
STF

Cingulate NEP and SVZ
intermingled with the callosal GEP

Parietal NEP and SVZ

STF1 t2

STF2

STF5

STF6

PARIETAL LOBE

Interhemispheric
fissure

Lateral parietal (granular) STF

STF1 t2

STF2

STF3a

STF3b

STF3c

STF4 t2

STF5?

STF6

Uncrossed
callosal
fibers

Callosal
GEP

Choroid
plexus

Cingulum

Cingulate
gyrus

Cortical/striatal
NEP and SVZ

Lateral
ventricle

Corpus
callosum

Subgranular
zone

Fornical
GEP

Dorsal
hippocampus

Habenula

Posterior striatal
NEP and SVZ

Layer I

Epithalamic
G/EP

Cortical plate

Stria
medullaris

Fornix

Dorsal complex

Thalamic G/EP

Layer VII
(subplate)

Ventral complex

THALAMUS

Strionuclear
GEP

Central
complex

Periventricular complex

Third ventricle

Subpial granular
layer (GEP)

Reticular
nucleus

Perifascicular GEP
(surrounds fiber tracts)

Caudate
nucleus
(tail)

Internal capsule
(posterior limb)

Stria
terminalis

Optic
tract

Zona incerta

Mammillo-
thalamic
tract

Subthalamic
G/EP

Amygdaloid
NEP

Subthalamic
nucleus

Cerebral peduncle

Forel's
fields

Medial
nucleus

STF1 t2

Caudate
nucleus
(tail)

Fornix

HYPO-
THALAMUS

Lateral
ventricle

STF2

AMYGDALA

Mammillary
body

Hypothalamic
G/EP

STF4 t2

TEMPORAL
LOBE

Basal
nucleus

Cortical
nucleus

Ventral
hippocampus

STF5

Amygdalo-
hippocampal
area?

Third ventricle
(mammillary recess)

Temporal (granular)
STF

Entorhinal cortex

Parahippocampal
gyrus

Temporal NEP and SVZ

Parahippocampal STF

Parahippocampal NEP and SVZ

G/EP - glioepithelium/ependyma
GEP - glioepithelium
NEP - neuroepithelium
SVZ - subventricular zone

Germinal and transitional structures in *italics*

PLATE 42A
CR 150 mm, GW 17, Y15-63
Frontal
Section 621

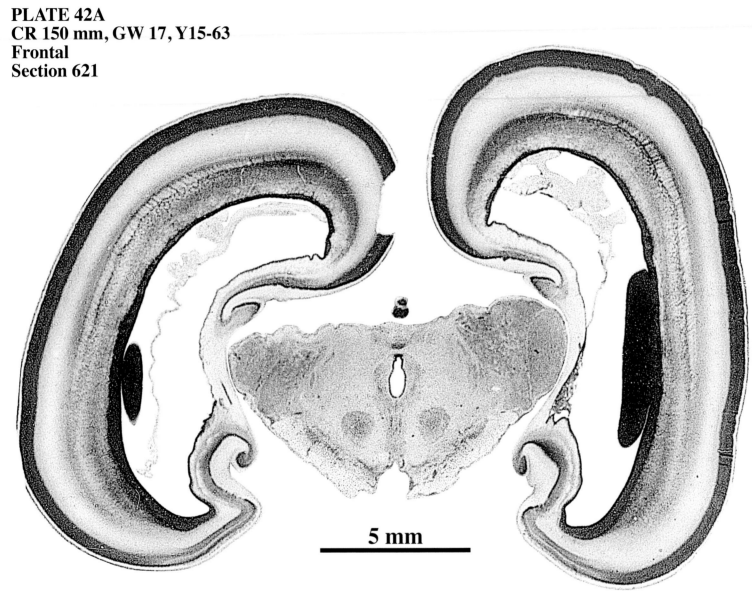

5 mm

LAYERS OF THE
CORTICAL
STRATIFIED
TRANSITIONAL
FIELD (STF)

STF1—Superficial fibrous layer with an early developmental stage *(t1)* when many cells are migrating through it, followed by a late stage *(t2)* with sparse cells. Endures as the subcortical white matter.

STF2—Upper cellular layer, the last sojourn zone before cells translocate to the cortical plate.

STF3—Honeycomb trilaminar matrix *(3a, 3b, 3c)* of cells and fibers found only in granular cortices.

STF4—Complex middle layer with three developmental stages: *t1*– fibrous layer without interspersed cells; *t2*– cells and fibers intermingle to form striations; *t3*– fibers endure in the deep white matter.

STF5—Deep cellular layer, the first sojourn zone to appear outside the germinal matrix.

STF6—Late-forming deep layer of callosal fibers outside the germinal matrix.

See detail of brain core in Plates 56A and B.

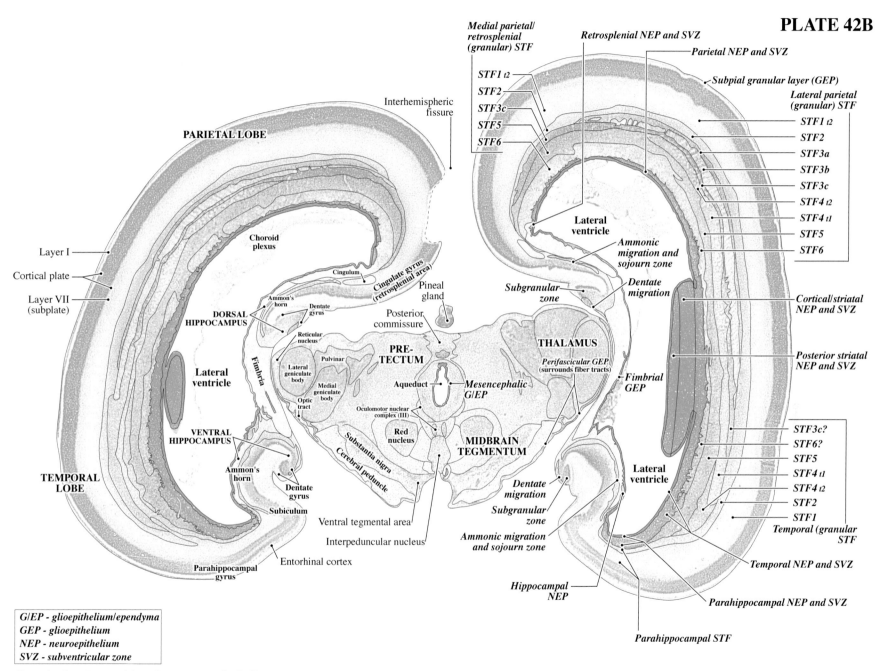

Interhemispheric fissure

Medial parietal/ retrosplenial (granular) STF

Retrosplenial NEP and SVZ

Parietal NEP and SVZ

STF1 t2
STF2
STF3c
STF5
STF6

Subpial granular layer (GEP)

Lateral parietal (granular) STF

STF1 t2
STF2
STF3a
STF3b
STF3c
STF4 t2
STF4 t1
STF5
STF6

PARIETAL LOBE

Layer I

Cortical plate

Layer VII (subplate)

Choroid plexus

Cingulum

Cingulate gyrus (retrosplenial area)

Pineal gland

Posterior commissure

Ammon's horn

Dentate gyrus

DORSAL HIPPOCAMPUS

Reticular nucleus

Fimbria

Lateral geniculate body

Pulvinar

Medial geniculate body

Optic tract

Lateral ventricle

PRE-TECTUM

Aqueduct

Mesencephalic G/EP

THALAMUS

Perifascicular GEP (surrounds fiber tracts)

Lateral ventricle

Ammonic migration and sojourn zone

Dentate migration

Subgranular zone

Cortical/striatal NEP and SVZ

Posterior striatal NEP and SVZ

Fimbrial GEP

VENTRAL HIPPOCAMPUS

Ammon's horn

Dentate gyrus

Subiculum

Oculomotor nuclear complex (III)

Red nucleus

MIDBRAIN TEGMENTUM

Substantia nigra

Cerebral peduncle

TEMPORAL LOBE

Ventral tegmental area

Interpeduncular nucleus

Dentate migration

Subgranular zone

Ammonic migration and sojourn zone

Lateral ventricle

STF3c?
STF6?
STF5
STF4 t1
STF4 t2
STF2
STF1

Temporal (granular STF

Temporal NEP and SVZ

Parahippocampal NEP and SVZ

Hippocampal NEP

Parahippocampal gyrus

Entorhinal cortex

Parahippocampal STF

G/EP - glioepithelium/ependyma
GEP - glioepithelium
NEP - neuroepithelium
SVZ - subventricular zone

Germinal and transitional structures in *italics*

**PLATE 43A
CR 150 mm, GW 17, Y15-63
Frontal
Section 681**

See this area of cortex from Section 711
in Plate 66.

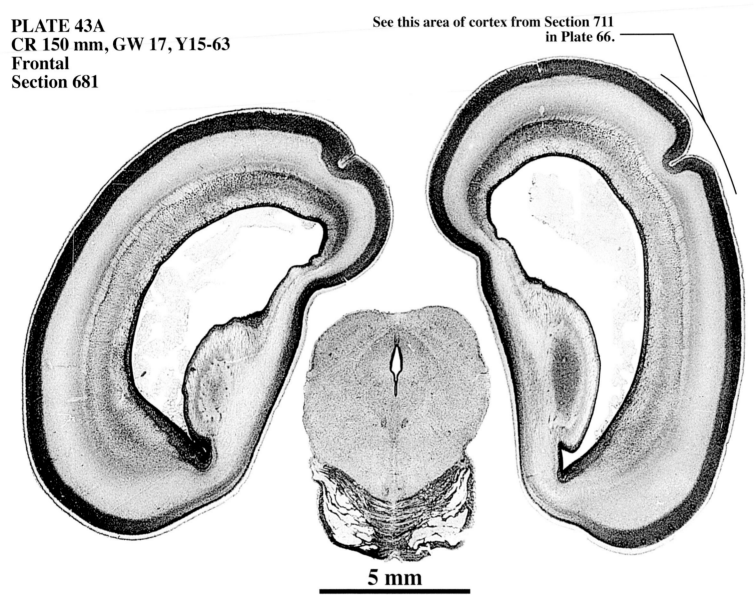

5 mm

LAYERS OF THE
CORTICAL
*STRATIFIED
TRANSITIONAL
FIELD (STF)*

STF1—Superficial
fibrous layer with an
early developmental
stage *(t1)* when many
cells are migrating
through it, followed by a
late stage *(t2)* with sparse
cells. Endures as the
subcortical white matter.

STF2—Upper cellular
layer, the last sojourn
zone before cells
translocate to the cortical
plate.

STF3—Honeycomb
trilaminar matrix *(3a, 3b,
3c)* of cells and fibers
found only in granular
cortices.

STF4—Complex middle
layer with three
developmental stages:
t1– fibrous layer without
interspersed cells;
t2– cells and fibers
intermingle to form
striations; *t3*– fibers
endure in the deep white
matter.

STF5—Deep cellular
layer, the first sojourn
zone to appear outside
the germinal matrix.

STF6—Late-forming
deep layer of callosal
fibers outside the
germinal matrix.

See detail of brain core in Plates 57A and B.

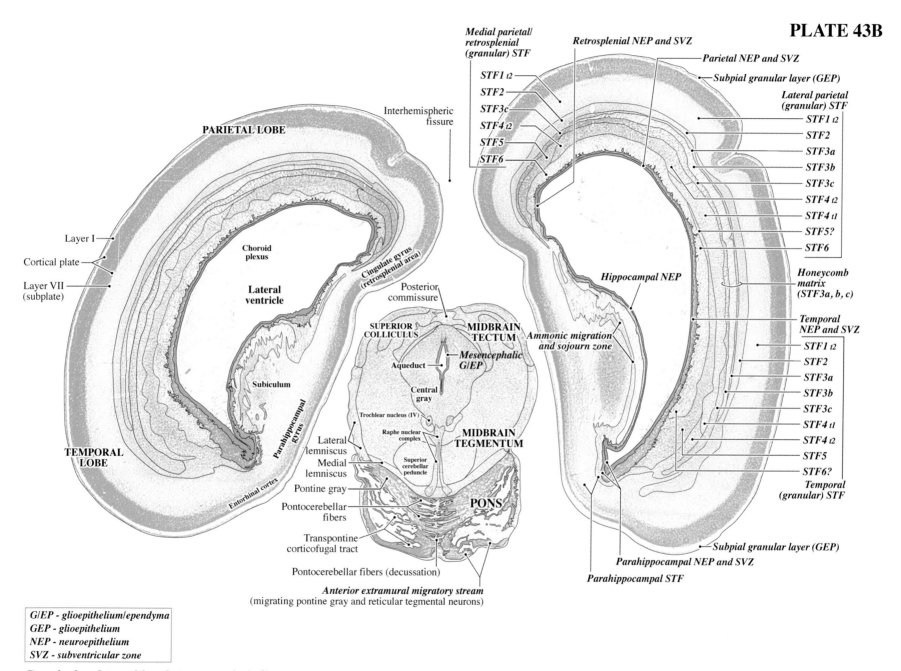

Medial parietal/
retrosplenial
(granular) STF

Retrosplenial NEP and SVZ

Parietal NEP and SVZ

Subpial granular layer (GEP)

STF1 t2
STF2
STF3c
STF4 t2
STF5
STF6

Lateral parietal
(granular) STF

STF1 t2
STF2
STF3a
STF3b
STF3c
STF4 t2
STF4 t1
STF5?
STF6

Honeycomb
matrix
(STF3a, b, c)

Hippocampal NEP

Temporal
NEP and SVZ

STF1 t2
STF2
STF3a
STF3b
STF3c
STF4 t1
STF4 t2
STF5
STF6?

Temporal
(granular) STF

Subpial granular layer (GEP)

Parahippocampal NEP and SVZ

Parahippocampal STF

Interhemispheric
fissure

PARIETAL LOBE

Layer I

Cortical plate

Layer VII
(subplate)

Choroid
plexus

Lateral
ventricle

Cingulate gyrus
(retrosplenial area)

Posterior
commissure

SUPERIOR
COLLICULUS

MIDBRAIN
TECTUM

Aqueduct

Mesencephalic
G/EP

Central
gray

Ammonic migration
and sojourn zone

Subiculum

Parahippocampal gyrus

TEMPORAL
LOBE

Entorhinal cortex

Trochlear nucleus (IV)

Raphe nuclear
complex

MIDBRAIN
TEGMENTUM

Lateral
lemniscus

Medial
lemniscus

Pontine gray

Pontocerebellar
fibers

Transpontine
corticofugal tract

Superior
cerebellar
peduncle

PONS

Pontocerebellar fibers (decussation)

Anterior extramural migratory stream
(migrating pontine gray and reticular tegmental neurons)

G/EP - glioepithelium/ependyma
GEP - glioepithelium
NEP - neuroepithelium
SVZ - subventricular zone

Germinal and transitional structures in italics

PLATE 44A
CR 150 mm, GW 17, Y15-63
Frontal
Section 741

See this area of cortex
from Section 711
in Plate 66.

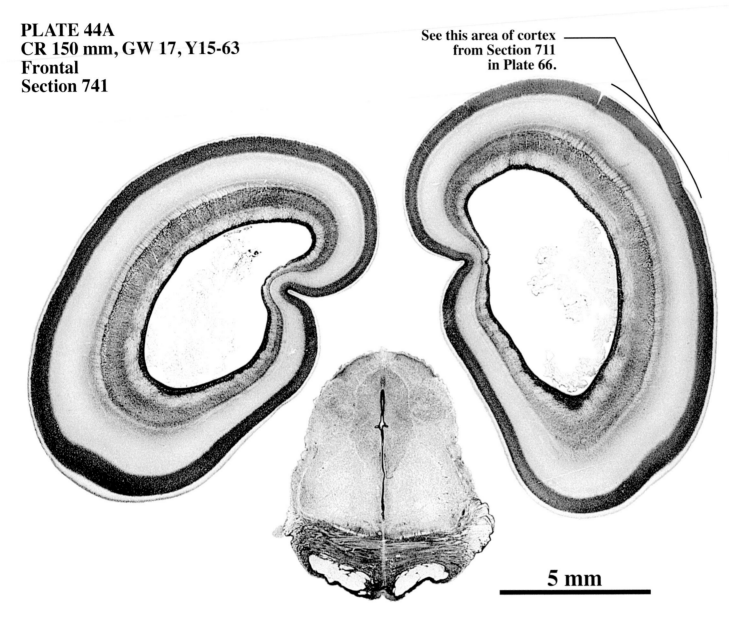

LAYERS OF THE
CORTICAL *STRATIFIED*
TRANSITIONAL FIELD (STF)

STF1—Superficial fibrous
layer with an early
developmental stage *(t1)* when
many cells are migrating
through it, followed by a late
stage *(t2)* with sparse cells.
Endures as the subcortical
white matter.

STF2—Upper cellular layer,
the last sojourn zone before
cells translocate to the cortical
plate.

STF3—Honeycomb trilaminar
matrix *(3a, 3b, 3c)* of cells and
fibers found only in granular
cortices.

STF4—Complex middle layer
with three developmental
stages:
t1– fibrous layer without
interspersed cells;
t2– cells and fibers intermingle
to form striations; *t3*– fibers
endure in the deep white
matter.

STF5—Deep cellular layer,
the first sojourn zone to appear
outside the germinal matrix.

STF6—Late-forming deep
layer of callosal fibers outside
the germinal matrix.

5 mm

See detail of brain core in Plates 58A and B.

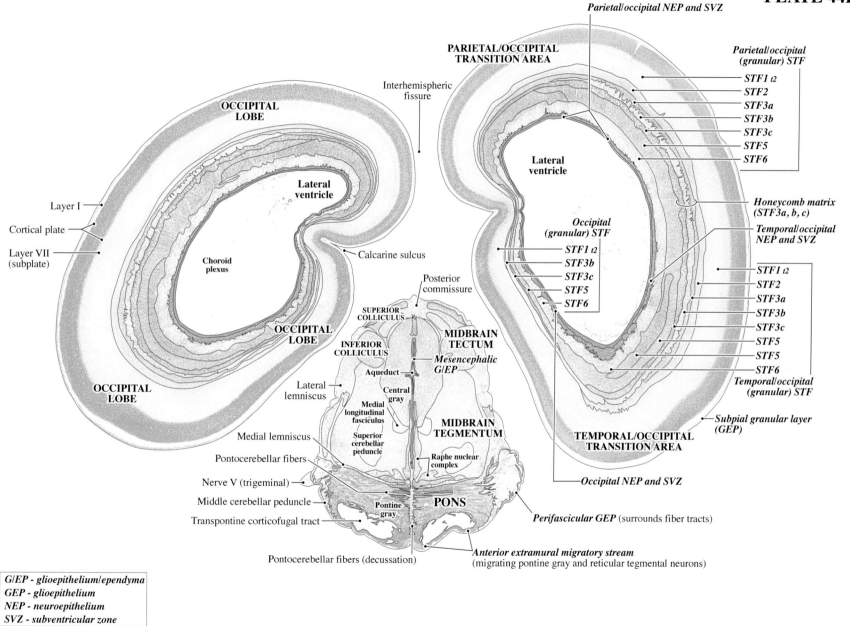

Parietal/occipital NEP and SVZ

**PARIETAL/OCCIPITAL
TRANSITION AREA**

*Parietal/occipital
(granular) STF*

Interhemispheric
fissure

**OCCIPITAL
LOBE**

STF1 t2
STF2
STF3a
STF3b
STF3c
STF5
STF6

Lateral
ventricle

Lateral
ventricle

Layer I

Cortical plate

Layer VII
(subplate)

*Honeycomb matrix
(STF3a, b, c)*

*Temporal/occipital
NEP and SVZ*

Choroid
plexus

*Occipital
(granular) STF*

Calcarine sulcus

**OCCIPITAL
LOBE**

STF1 t2
STF3b
STF3c
STF5
STF6

STF1 t2
STF2
STF3a
STF3b
STF3c
STF5
STF5
STF6

Posterior
commissure

**SUPERIOR
COLLICULUS**

**MIDBRAIN
TECTUM**

**OCCIPITAL
LOBE**

**INFERIOR
COLLICULUS**

Aqueduct

*Mesencephalic
G/EP*

*Temporal/occipital
(granular) STF*

Lateral
lemniscus

Central
gray

**MIDBRAIN
TEGMENTUM**

Medial
longitudinal
fasciculus

*Subpial granular layer
(GEP)*

Medial lemniscus

Superior
cerebellar
peduncle

**TEMPORAL/OCCIPITAL
TRANSITION AREA**

Pontocerebellar fibers

*Raphe nuclear
complex*

Nerve V (trigeminal)

Occipital NEP and SVZ

Middle cerebellar peduncle

Pontine
gray

PONS

Transpontine corticofugal tract

Perifascicular GEP (surrounds fiber tracts)

Pontocerebellar fibers (decussation)

Anterior extramural migratory stream
(migrating pontine gray and reticular tegmental neurons)

G/EP - glioepithelium/ependyma
GEP - glioepithelium
NEP - neuroepithelium
SVZ - subventricular zone

Germinal and transitional structures in *italics*

PLATE 45A
CR 150 mm, GW 17, Y15-63
Frontal
Section 801

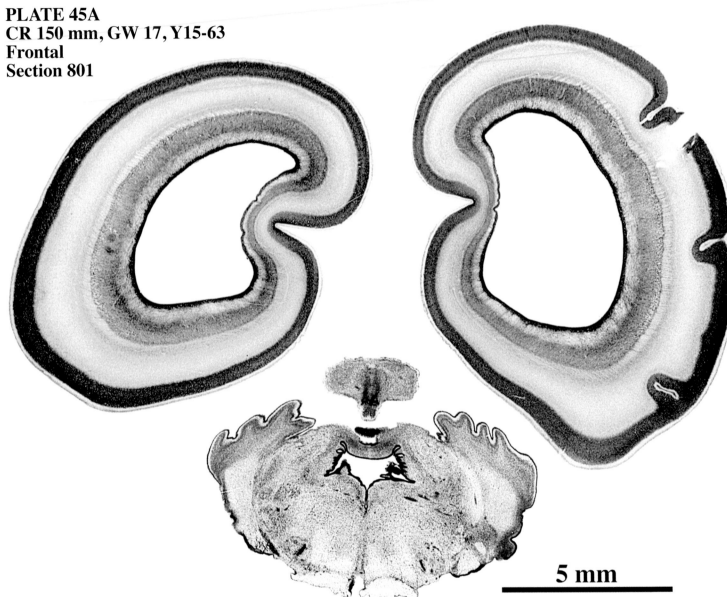

5 mm

See detail of brain core and cerebellum in Plates 59A and B.

LAYERS OF THE CORTICAL *STRATIFIED TRANSITIONAL FIELD (STF)*

STF1—Superficial fibrous layer with an early developmental stage *(t1)* when many cells are migrating through it, followed by a late stage *(t2)* with sparse cells. Endures as the subcortical white matter.

STF2—Upper cellular layer, the last sojourn zone before cells translocate to the cortical plate.

STF3—Honeycomb trilaminar matrix *(3a, 3b, 3c)* of cells and fibers found only in granular cortices.

STF4—Complex middle layer with three developmental stages: *t1*– fibrous layer without interspersed cells; *t2*– cells and fibers intermingle to form striations; *t3*– fibers endure in the deep white matter.

STF5—Deep cellular layer, the first sojourn zone to appear outside the germinal matrix.

STF6—Late-forming deep layer of callosal fibers outside the germinal matrix.

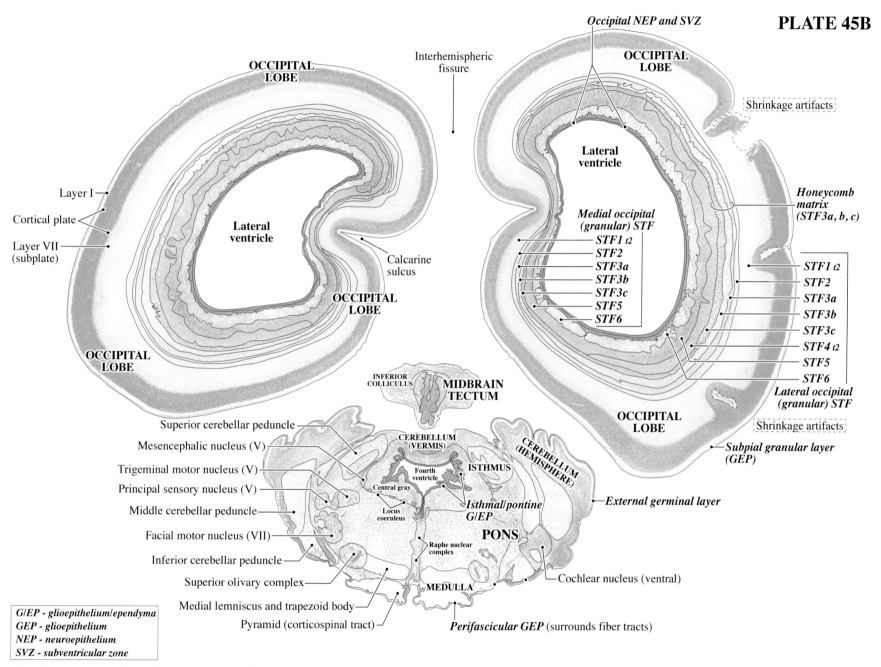

Occipital NEP and SVZ

Interhemispheric fissure

OCCIPITAL LOBE

OCCIPITAL LOBE

Shrinkage artifacts

Layer I

Cortical plate

Lateral ventricle

Layer VII (subplate)

Calcarine sulcus

OCCIPITAL LOBE

OCCIPITAL LOBE

Lateral ventricle

Medial occipital (granular) STF

STF1 t2
STF2
STF3a
STF3b
STF3c
STF5
STF6

Honeycomb matrix (STF3a, b, c)

STF1 t2
STF2
STF3a
STF3b
STF3c
STF4 t2
STF5
STF6

Lateral occipital (granular) STF

OCCIPITAL LOBE

Shrinkage artifacts

Subpial granular layer (GEP)

INFERIOR COLLICULUS

MIDBRAIN TECTUM

Superior cerebellar peduncle

Mesencephalic nucleus (V)

Trigeminal motor nucleus (V)

Principal sensory nucleus (V)

Middle cerebellar peduncle

Facial motor nucleus (VII)

Inferior cerebellar peduncle

Superior olivary complex

Medial lemniscus and trapezoid body

Pyramid (corticospinal tract)

CEREBELLUM (VERMIS)

CEREBELLUM (HEMISPHERE)

ISTHMUS

Fourth ventricle

Central gray

Locus coeruleus

Isthmal/pontine G/EP

PONS

Raphe nuclear complex

MEDULLA

External germinal layer

Cochlear nucleus (ventral)

Perifascicular GEP (surrounds fiber tracts)

G/EP - glioepithelium/ependyma
GEP - glioepithelium
NEP - neuroepithelium
SVZ - subventricular zone

Germinal and transitional structures in *italics*

PLATE 46A
CR 150 mm, GW 17, Y15-63
Frontal
Section 821

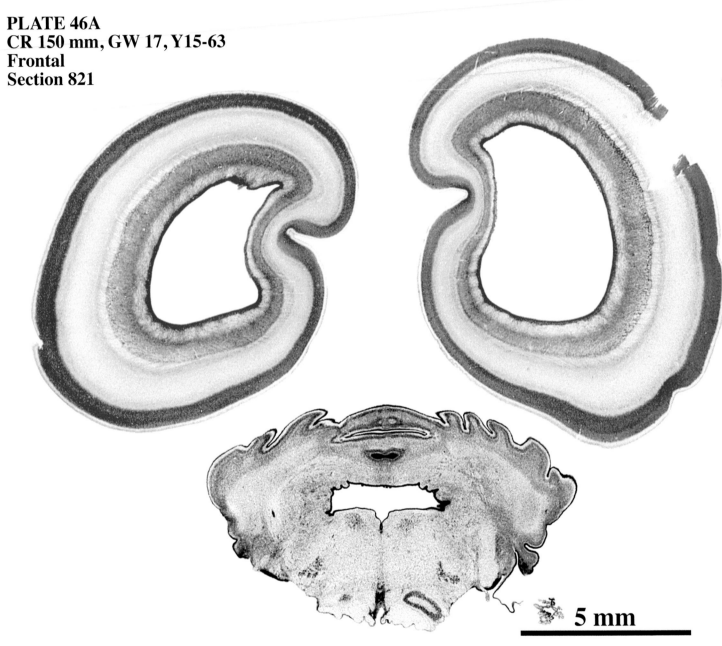

5 mm

See detail of brain core and cerebellum in Plates 60A and B.

LAYERS OF THE CORTICAL *STRATIFIED TRANSITIONAL FIELD (STF)*

STF1 —Superficial fibrous layer with an early developmental stage *(t1)* when many cells are migrating through it, followed by a late stage *(t2)* with sparse cells. Endures as the subcortical white matter.

STF2 —Upper cellular layer, the last sojourn zone before cells translocate to the cortical plate.

STF3 —Honeycomb trilaminar matrix *(3a, 3b, 3c)* of cells and fibers found only in granular cortices.

STF4 —Complex middle layer with three developmental stages: *t1* – fibrous layer without interspersed cells; *t2* – cells and fibers intermingle to form striations; *t3* – fibers endure in the deep white matter.

STF5 —Deep cellular layer, the first sojourn zone to appear outside the germinal matrix.

STF6 —Late-forming deep layer of callosal fibers outside the germinal matrix.

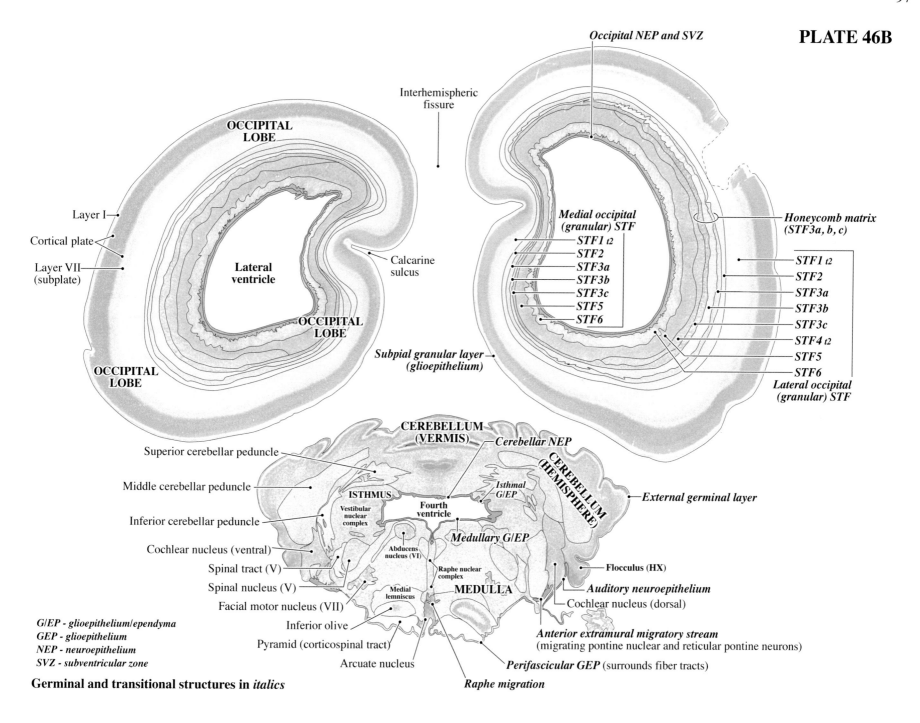

Occipital NEP and SVZ

Interhemispheric
fissure

OCCIPITAL
LOBE

*Medial occipital
(granular) STF*

*Honeycomb matrix
(STF3a, b, c)*

Layer I

Cortical plate

STF1 t2
STF2
STF3a
STF3b
STF3c
STF5
STF6

Layer VII
(subplate)

Lateral
ventricle

Calcarine
sulcus

STF1 t2
STF2
STF3a
STF3b
STF3c
STF4 t2
STF5
STF6

OCCIPITAL
LOBE

OCCIPITAL
LOBE

*Subpial granular layer
(glioepithelium)*

*Lateral occipital
(granular) STF*

CEREBELLUM
(VERMIS)

Cerebellar NEP

CEREBELLUM
(HEMISPHERE)

Superior cerebellar peduncle

ISTHMUS

*Isthmal
G/EP*

Middle cerebellar peduncle

Vestibular
nuclear
complex

Fourth
ventricle

External germinal layer

Inferior cerebellar peduncle

Abducens
nucleus (VI)

Medullary G/EP

Cochlear nucleus (ventral)

Raphe nuclear
complex

Spinal tract (V)

Flocculus (HX)

Spinal nucleus (V)

Medial
lemniscus

MEDULLA

Auditory neuroepithelium

Facial motor nucleus (VII)

Cochlear nucleus (dorsal)

G/EP - glioepithelium/ependyma
GEP - glioepithelium
NEP - neuroepithelium
SVZ - subventricular zone

Inferior olive

Anterior extramural migratory stream
(migrating pontine nuclear and reticular pontine neurons)

Pyramid (corticospinal tract)

Perifascicular GEP (surrounds fiber tracts)

Arcuate nucleus

Germinal and transitional structures in *italics*

Raphe migration

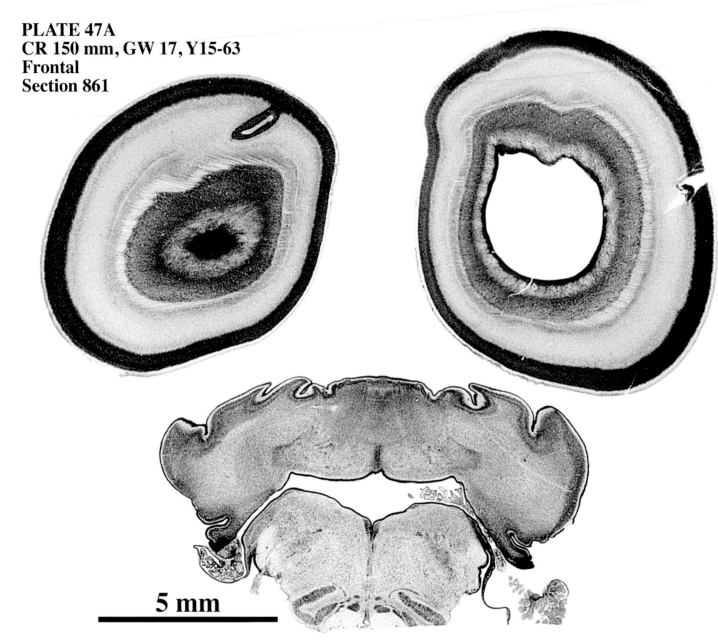

PLATE 47A
CR 150 mm, GW 17, Y15-63
Frontal
Section 861

5 mm

See detail of brain core and cerebellum in Plates 61A and B.

**LAYERS OF THE
CORTICAL *STRATIFIED
TRANSITIONAL FIELD (STF)***

STF1—Superficial fibrous
layer with an early
developmental stage *(t1)*
when many cells are
migrating through it,
followed by a late stage *(t2)*
with sparse cells. Endures as
the subcortical white matter.

STF2—Upper cellular layer,
the last sojourn zone before
cells translocate to the
cortical plate.

STF3—Honeycomb
trilaminar matrix *(3a, 3b, 3c)*
of cells and fibers found only
in granular cortices.

STF4—Complex middle
layer with three
developmental stages:
t1– fibrous layer without
interspersed cells;
t2– cells and fibers
intermingle to form
striations; *t3*– fibers endure
in the deep white matter.

STF5—Deep cellular layer,
the first sojourn zone to
appear outside the germinal
matrix.

STF6—Late-forming deep
layer of callosal fibers
outside the germinal matrix.

PLATE 47B

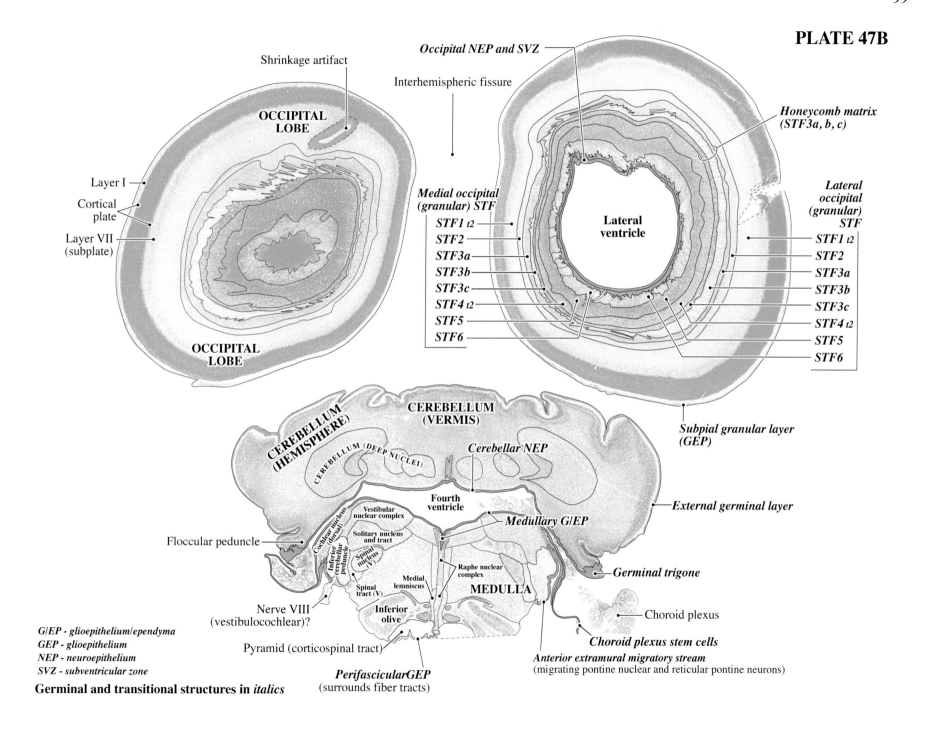

Shrinkage artifact

Occipital NEP and SVZ

Interhemispheric fissure

OCCIPITAL LOBE

Honeycomb matrix (STF3a, b, c)

Layer I

Cortical plate

Layer VII (subplate)

OCCIPITAL LOBE

Medial occipital (granular) STF

STF1 t2
STF2
STF3a
STF3b
STF3c
STF4 t2
STF5
STF6

Lateral ventricle

Lateral occipital (granular) STF

STF1 t2
STF2
STF3a
STF3b
STF3c
STF4 t2
STF5
STF6

CEREBELLUM (VERMIS)

CEREBELLUM (HEMISPHERE)

CEREBELLUM (DEEP NUCLEI)

Cerebellar NEP

Subpial granular layer (GEP)

Fourth ventricle

Medullary G/EP

External germinal layer

Vestibular nuclear complex

Cochlear nucleus (dorsal)

Flocular peduncle

Solitary nucleus and tract

Inferior cerebellar peduncle

Spinal nucleus (V)

Raphe nuclear complex

Germinal trigone

Medial lemniscus

MEDULLA

Spinal tract (V)

Nerve VIII (vestibulocochlear)?

Medial lemniscus

Inferior olive

Choroid plexus

Choroid plexus stem cells

Pyramid (corticospinal tract)

PerifascicularGEP (surrounds fiber tracts)

Anterior extramural migratory stream (migrating pontine nuclear and reticular pontine neurons)

G/EP - glioepithelium/ependyma
GEP - glioepithelium
NEP - neuroepithelium
SVZ - subventricular zone

Germinal and transitional structures in *italics*

PLATE 48A
CR 150 mm, GW 17, Y15-63
Frontal
Section 911

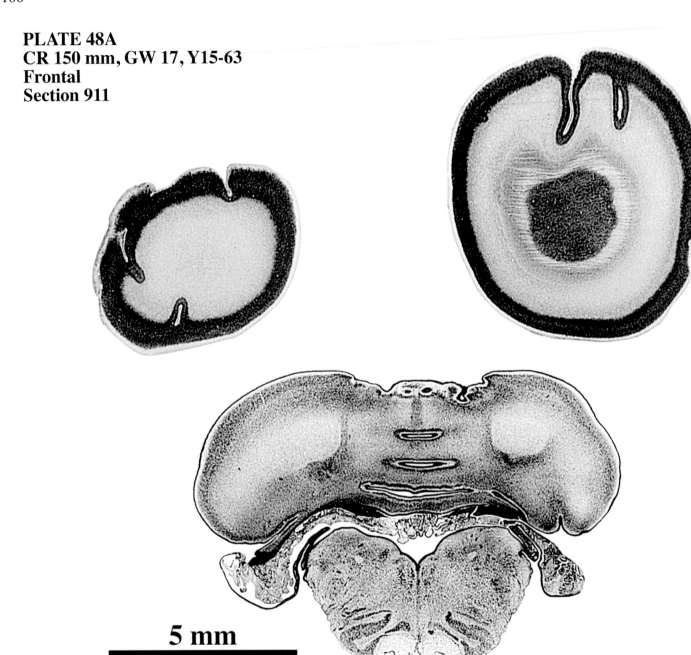

5 mm

LAYERS OF THE
CORTICAL *STRATIFIED*
TRANSITIONAL FIELD (STF)

STF1—Superficial fibrous layer with an early developmental stage *(t1)* when many cells are migrating through it, followed by a late stage *(t2)* with sparse cells. Endures as the subcortical white matter.

STF2—Upper cellular layer, the last sojourn zone before cells translocate to the cortical plate.

STF3—Honeycomb trilaminar matrix *(3a, 3b, 3c)* of cells and fibers found only in granular cortices.

STF4—Complex middle layer with three developmental stages:
t1– fibrous layer without interspersed cells;
t2– cells and fibers intermingle to form striations; *t3*– fibers endure in the deep white matter.

STF5—Deep cellular layer, the first sojourn zone to appear outside the germinal matrix.

See detail of the brain core and cerebellum in Plates 62A and B.

PLATE 48B

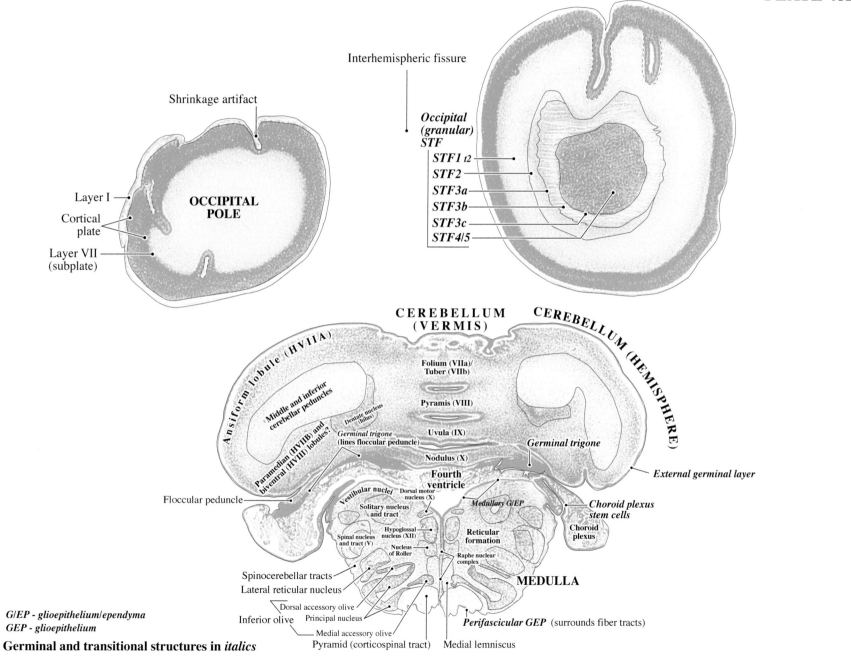

Interhemispheric fissure

Shrinkage artifact

Occipital (granular) STF

STF1 t2
STF2
STF3a
STF3b
STF3c
STF4/5

Layer I

OCCIPITAL POLE

Cortical plate

Layer VII (subplate)

CEREBELLUM (VERMIS)

CEREBELLUM (HEMISPHERE)

Ansiform lobule (HVIIA)

Middle and inferior cerebellar peduncles

Dentate nucleus (hilus)

Germinal trigone (lines floccular peduncle)

Paramedian (HVIIB) and biventral (HVIII) lobules?

Folium (VIIa)/ Tuber (VIIb)

Pyramis (VIII)

Uvula (IX)

Nodulus (X)

Germinal trigone

External germinal layer

Fourth ventricle

Floccular peduncle

Vestibular nuclei

Dorsal motor nucleus (X)

Solitary nucleus and tract

Medullary G/EP

Choroid plexus stem cells

Hypoglossal nucleus (XII)

Reticular formation

Choroid plexus

Spinal nucleus and tract (V)

Nucleus of Roller

Raphe nuclear complex

MEDULLA

Spinocerebellar tracts

Lateral reticular nucleus

G/EP - glioepithelium/ependyma
GEP - glioepithelium

Germinal and transitional structures in *italics*

Dorsal accessory olive

Inferior olive

Principal nucleus

Medial accessory olive

Pyramid (corticospinal tract)

Medial lemniscus

Perifascicular GEP (surrounds fiber tracts)

PLATE 49A
CR 150 mm, GW 17, Y15-63
Frontal
Section 961

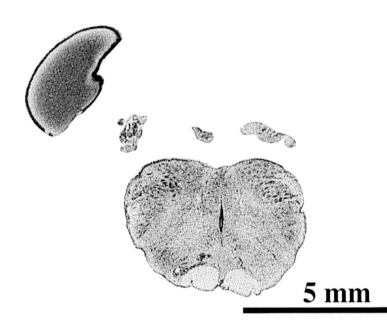

5 mm

PLATE 49B

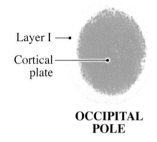

Layer I →

Cortical
plate

**OCCIPITAL
POLE**

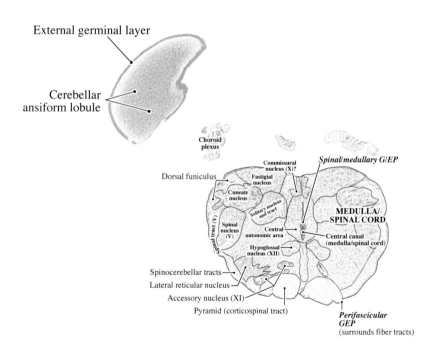

External germinal layer

Cerebellar
ansiform lobule

Choroid
plexus

Commissural
nucleus (X)?

Spinal/medullary G/EP

Dorsal funiculus

Fastigial
nucleus

Cuneate
nucleus

Solitary nucleus
and tract

**MEDULLA/
SPINAL CORD**

Spinal
tract (V)

Spinal
nucleus
(V)

Central
autonomic area

Central canal
(medulla/spinal cord)

Hypoglossal
nucleus (XII)

Spinocerebellar tracts

Lateral reticular nucleus

Accessory nucleus (XI)

Pyramid (corticospinal tract)

*Perifascicular
GEP*
(surrounds fiber tracts)

G/EP - glioepithelium/ependyma
GEP - glioepithelium
Germinal and transitional structures in *italics*

PLATE 50A
CR 150 mm, GW 17, Y15-63
Frontal
Section 341

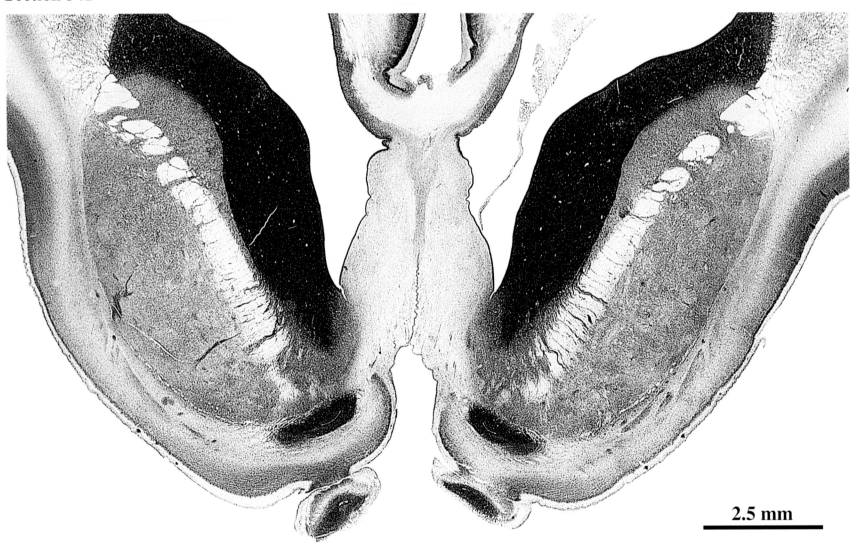

2.5 mm

See the entire Section in Plates 36A and B.

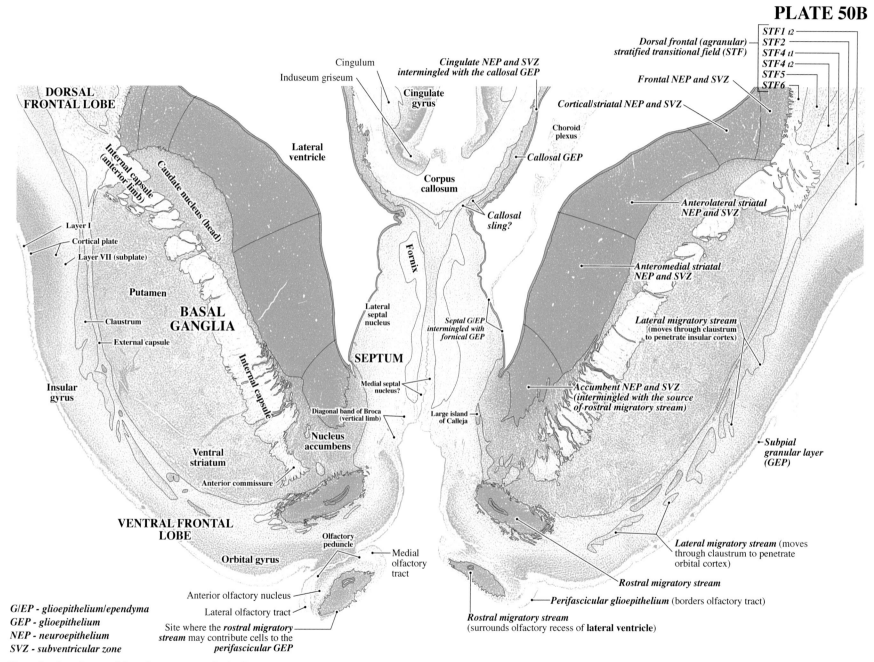

PLATE 50B

STF1 *t2*
STF2
STF4 *t1*
STF4 *t2*
STF5
STF6

Dorsal frontal (agranular)
stratified transitional field (STF)

Frontal NEP and SVZ

Cortical/striatal NEP and SVZ

Anterolateral striatal
NEP and SVZ

Anteromedial striatal
NEP and SVZ

Lateral migratory stream
(moves through claustrum
to penetrate insular cortex)

Accumbent NEP and SVZ
(intermingled with the source
of rostral migratory stream)

Subpial
granular layer
(GEP)

Lateral migratory stream (moves
through claustrum to penetrate
orbital cortex)

Rostral migratory stream

Perifascicular glioepithelium (borders olfactory tract)

Rostral migratory stream
(surrounds olfactory recess of **lateral ventricle**)

Cingulum

Induseum griseum

Cingulate NEP and SVZ
intermingled with the callosal GEP

Cingulate
gyrus

Choroid
plexus

Callosal GEP

Corpus
callosum

Callosal
sling?

Fornix

Lateral
septal
nucleus

Septal G/EP
intermingled with
fornical GEP

SEPTUM

Medial septal
nucleus?

Large island
of Calleja

DORSAL
FRONTAL LOBE

Internal capsule
(anterior limb)

Caudate nucleus (head)

Lateral
ventricle

Layer I

Cortical plate

Layer VII (subplate)

Putamen

BASAL
GANGLIA

Claustrum

External capsule

Internal capsule

Insular
gyrus

Ventral
striatum

Diagonal band of Broca
(vertical limb)

Nucleus
accumbens

Anterior commissure

VENTRAL FRONTAL
LOBE

Orbital gyrus

Olfactory
peduncle

Medial
olfactory
tract

Anterior olfactory nucleus

Lateral olfactory tract

Site where the *rostral migratory*
stream may contribute cells to the
perifascicular GEP

G/EP - glioepithelium/ependyma
GEP - glioepithelium
NEP - neuroepithelium
SVZ - subventricular zone

Germinal and transitional structures in *italics*

PLATE 51A
CR 150 mm, GW 17, Y15-63
Frontal
Section 381

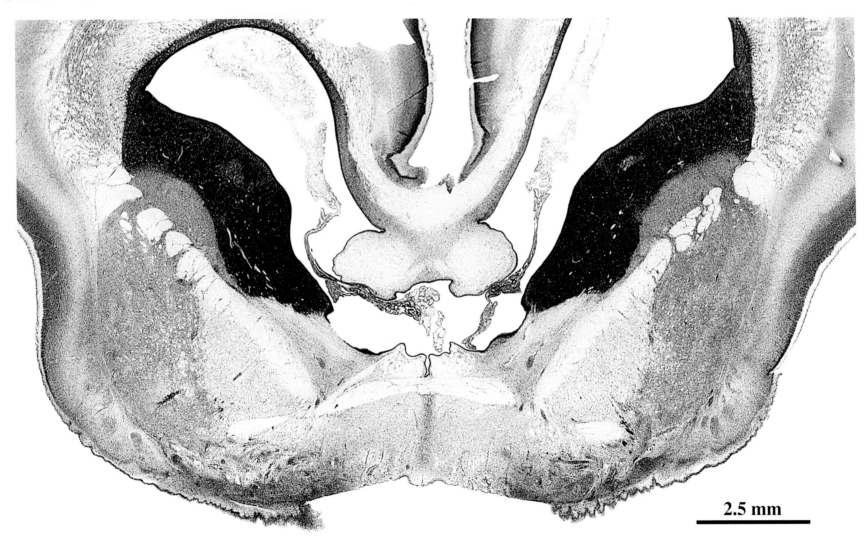

2.5 mm

See the entire Section in Plates 37A and B.

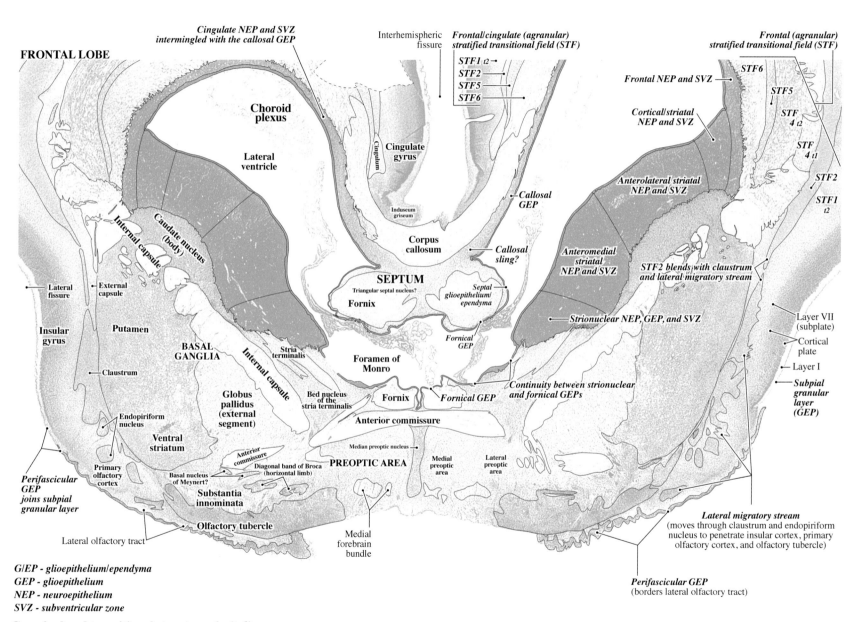

FRONTAL LOBE

Cingulate NEP and SVZ intermingled with the callosal GEP

Interhemispheric fissure

Frontal/cingulate (agranular) stratified transitional field (STF)

STF1 t2 →
STF2
STF5
STF6

Frontal (agranular) stratified transitional field (STF)

STF6

Frontal NEP and SVZ

STF5

STF 4 t2

STF 4 t1

STF2

STF1 t2

Choroid plexus

Cingulate gyrus

Cingulum

Lateral ventricle

Induseum griseum

Callosal GEP

Corpus callosum

Callosal sling?

Cortical/striatal NEP and SVZ

Anterolateral striatal NEP and SVZ

Caudate nucleus (body)

Internal capsule

Anteromedial striatal NEP and SVZ

STF2 blends with claustrum and lateral migratory stream

SEPTUM

Triangular septal nucleus?

Septal glioepithelium/ ependyma

Fornix

Lateral fissure

External capsule

Putamen

BASAL GANGLIA

Internal capsule

Stria terminalis

Fornical GEP

Insular gyrus

Claustrum

Globus pallidus (external segment)

Bed nucleus of the stria terminalis

Foramen of Monro

Fornix

Fornical GEP

Strionuclear NEP, GEP, and SVZ

Continuity between strionuclear and fornical GEPs

Layer VII (subplate)

Cortical plate

Layer I

Subpial granular layer (GEP)

Endopiriform nucleus

Ventral striatum

Anterior commissure

Anterior commissure

Median preoptic nucleus

PREOPTIC AREA

Medial preoptic area

Lateral preoptic area

Perifascicular GEP joins subpial granular layer

Primary olfactory cortex

Basal nucleus of Meynert?

Diagonal band of Broca (horizontal limb)

Substantia innominata

Olfactory tubercle

Lateral olfactory tract

Medial forebrain bundle

Lateral migratory stream (moves through claustrum and endopiriform nucleus to penetrate insular cortex, primary olfactory cortex, and olfactory tubercle)

Perifascicular GEP (borders lateral olfactory tract)

G/EP - glioepithelium/ependyma
GEP - glioepithelium
NEP - neuroepithelium
SVZ - subventricular zone

Germinal and transitional structures in *italics*

PLATE 52A
CR 150 mm, GW 17, Y15-63
Frontal
Section 441

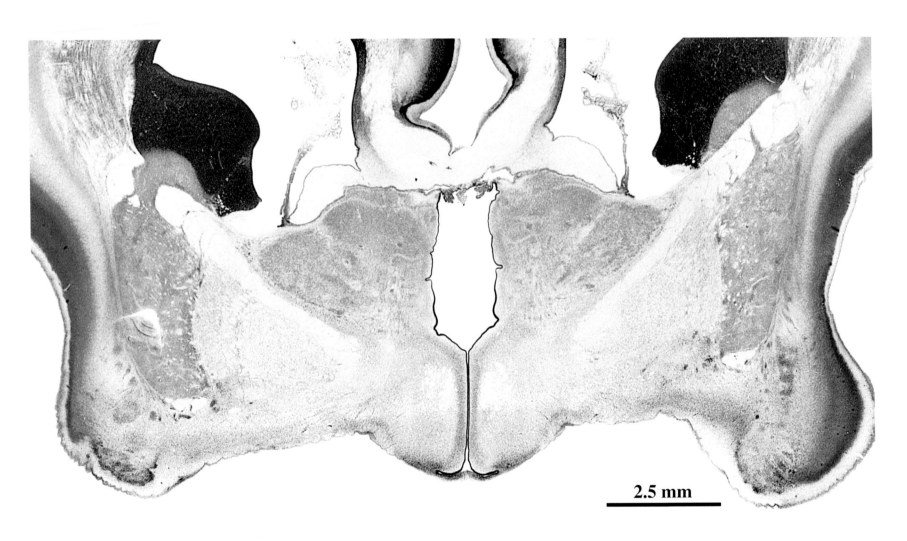

2.5 mm

See the entire Section in Plates 38A and B.

PLATE 52B

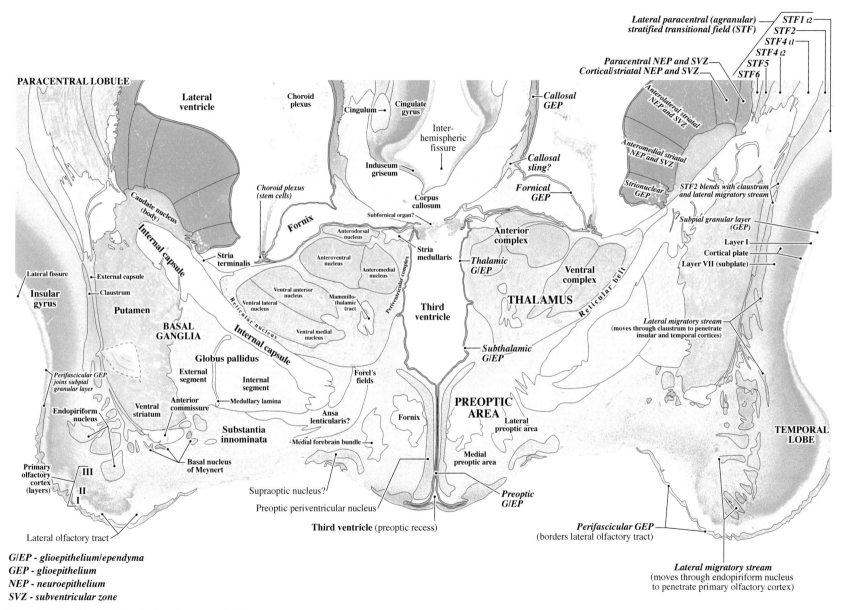

PARACENTRAL LOBULE

Lateral ventricle

Choroid plexus

Cingulum

Cingulate gyrus

Inter-hemispheric fissure

Callosal GEP

Lateral paracentral (agranular) stratified transitional field (STF)

STF1 t2
STF2
STF4 t1
STF4 t2
STF5
STF6

Paracentral NEP and SVZ
Cortical/striatal NEP and SVZ

Anterolateral striatal NEP and SVZ

Induseum griseum

Callosal sling?

Anteromedial striatal NEP and SVZ

Choroid plexus (stem cells)

Corpus callosum

Fornical GEP

Strionuclear GEP

STF2 blends with claustrum and lateral migratory stream

Caudate nucleus (body)

Fornix

Subfornical organ?

Anterodorsal nucleus

Stria medullaris

Anterior complex

Subpial granular layer (GEP)

Internal capsule

Stria terminalis

Anteroventral nucleus

Anteromedial nucleus

Thalamic G/EP

Ventral complex

Layer I

Cortical plate

Layer VII (subplate)

Lateral fissure

External capsule

Claustrum

Ventral anterior nucleus

Mammillo-thalamic tract

Ventral lateral nucleus

Periventricular complex

THALAMUS

Reticular belt

Insular gyrus

Putamen

BASAL GANGLIA

Reticular nucleus

Third ventricle

Lateral migratory stream (moves through claustrum to penetrate insular and temporal cortices)

Internal capsule

Ventral medial nucleus

Globus pallidus

External segment

Internal segment

Medullary lamina

Forel's fields

Subthalamic G/EP

Perifascicular GEP joins subpial granular layer

Anterior commissure

Ansa lenticularis?

Fornix

PREOPTIC AREA

Endopiriform nucleus

Ventral striatum

Lateral preoptic area

TEMPORAL LOBE

Substantia innominata

Medial forebrain bundle

Basal nucleus of Meynert

Medial preoptic area

Primary olfactory cortex (layers)

III

II

I

Supraoptic nucleus?

Preoptic periventricular nucleus

Preoptic G/EP

Lateral olfactory tract

Third ventricle (preoptic recess)

Perifascicular GEP (borders lateral olfactory tract)

Lateral migratory stream (moves through endopiriform nucleus to penetrate primary olfactory cortex)

G/EP - *glioepithelium/ependyma*
GEP - *glioepithelium*
NEP - *neuroepithelium*
SVZ - *subventricular zone*

Germinal and transitional structures in *italics*

PLATE 53A
CR 150 mm, GW 17, Y15-63
Frontal
Section 461

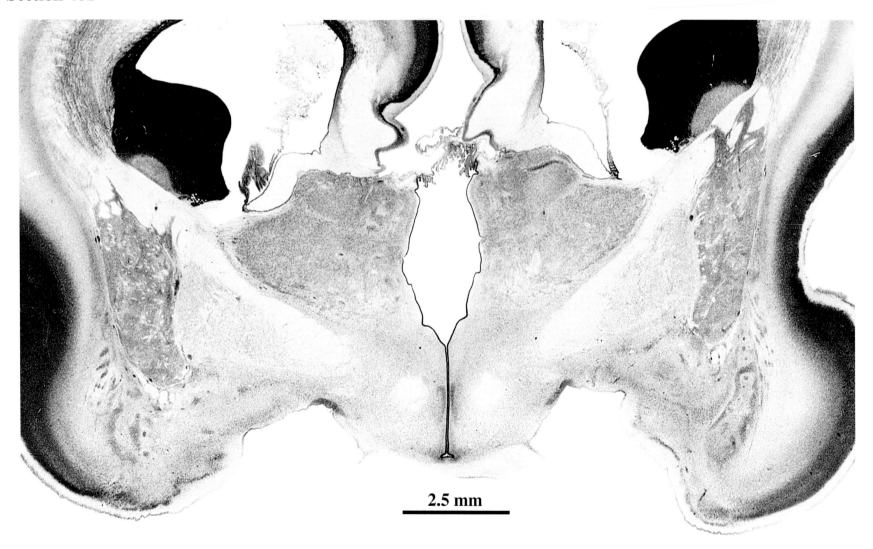

2.5 mm

See the entire Section in Plates 39A and B.

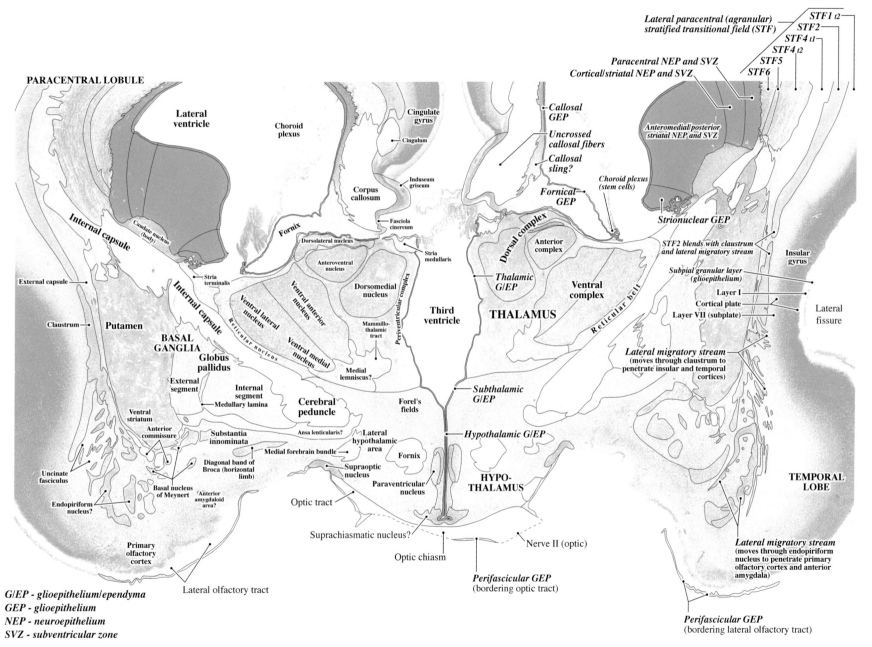

Lateral paracentral (agranular) stratified transitional field (STF)

STF1 t2
STF2
STF4 t1
STF4 t2
STF5
STF6

Paracentral NEP and SVZ
Cortical/striatal NEP and SVZ

Anteromedial/posterior striatal NEP and SVZ

PARACENTRAL LOBULE

Lateral ventricle

Choroid plexus

Cingulate gyrus

Callosal GEP

Uncrossed callosal fibers

Callosal sling?

Fornical GEP

Choroid plexus (stem cells)

Strionuclear GEP

Cingulum

Corpus callosum

Induseum griseum

Fasciola cinereum

Internal capsule

Caudate nucleus (body)

Fornix

Stria medullaris

Dorsolateral nucleus

Dorsal complex

Anterior complex

STF2 blends with claustrum and lateral migratory stream

Insular gyrus

External capsule

Stria terminalis

Anteroventral nucleus

Thalamic G/EP

Ventral complex

Subpial granular layer (glioepithelium)

Internal capsule

Ventral anterior nucleus

Dorsomedial nucleus

Layer I

Cortical plate

Layer VII (subplate)

Lateral fissure

Claustrum

Putamen

BASAL GANGLIA

Reticular nucleus

Ventral lateral nucleus

Ventral medial nucleus

Periventricular complex

Mammillo-thalamic tract

Third ventricle

THALAMUS

Reticular belt

Globus pallidus

External segment

Internal segment
Medullary lamina

Cerebral peduncle

Medial lemniscus?

Subthalamic G/EP

Lateral migratory stream (moves through claustrum to penetrate insular and temporal cortices)

Ventral striatum

Anterior commissure

Substantia innominata

Ansa lenticularis?

Lateral hypothalamic area

Forel's fields

Hypothalamic G/EP

Medial forebrain bundle

Diagonal band of Broca (horizontal limb)

Fornix

Uncinate fasciculus

Basal nucleus of Meynert

Anterior amygdaloid area?

Supraoptic nucleus

Paraventricular nucleus

HYPO-THALAMUS

TEMPORAL LOBE

Endopiriform nucleus?

Optic tract

Lateral migratory stream (moves through endopiriform nucleus to penetrate primary olfactory cortex and anterior amygdala)

Primary olfactory cortex

Suprachiasmatic nucleus?

Optic chiasm

Nerve II (optic)

Lateral olfactory tract

Perifascicular GEP (bordering optic tract)

G/EP - glioepithelium/ependyma
GEP - glioepithelium
NEP - neuroepithelium
SVZ - subventricular zone

Perifascicular GEP (bordering lateral olfactory tract)

Germinal and transitional structures in italics

PLATE 54A
CR 150 mm, GW 17, Y15-63
Frontal
Section 501

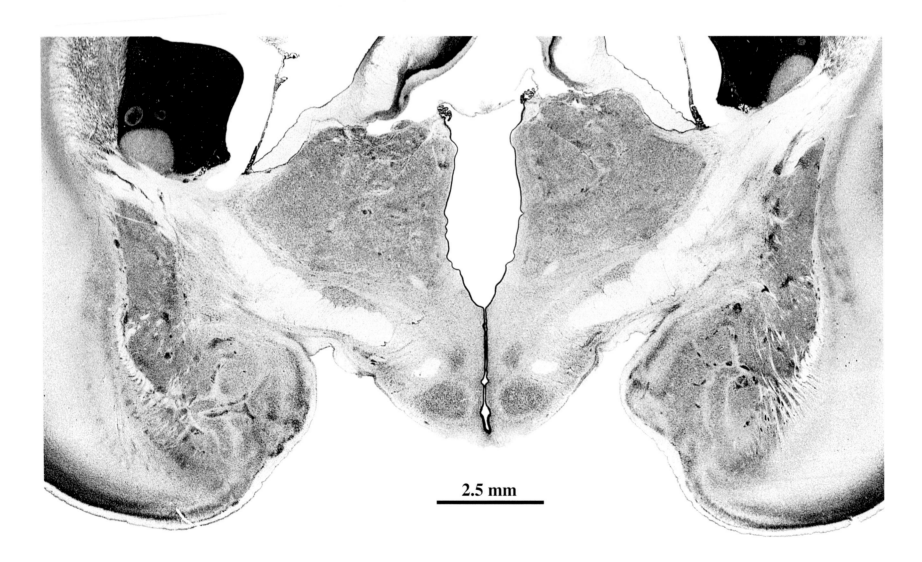

2.5 mm

See the entire Section in Plates 40A and B.

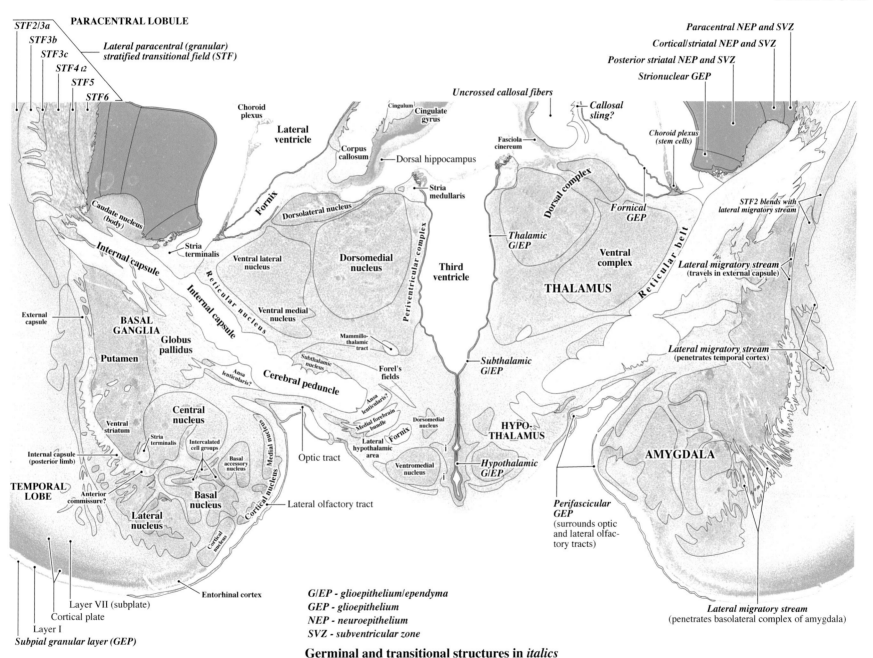

STF2/3a
STF3b
STF3c
STF4 t2
STF5
STF6

PARACENTRAL LOBULE

Lateral paracentral (granular) stratified transitional field (STF)

Choroid plexus

Lateral ventricle

Corpus callosum

Cingulum

Cingulate gyrus

Dorsal hippocampus

Uncrossed callosal fibers

Callosal sling?

Paracentral NEP and SVZ

Cortical/striatal NEP and SVZ

Posterior striatal NEP and SVZ

Strionuclear GEP

Choroid plexus (stem cells)

Fornix

Stria medullaris

Fasciola cinereum

Dorsolateral nucleus

Caudate nucleus (body)

Stria terminalis

Ventral lateral nucleus

Dorsomedial nucleus

Periventricular complex

Third ventricle

Dorsal complex

Fornical GEP

Ventral complex

THALAMUS

Reticular belt

STF2 blends with lateral migratory stream

Internal capsule

Thalamic G/EP

Reticular nucleus

Internal capsule

Ventral medial nucleus

Lateral migratory stream (travels in external capsule)

External capsule

BASAL GANGLIA

Globus pallidus

Putamen

Mammillo-thalamic tract

Subthalamic G/EP

Lateral migratory stream (penetrates temporal cortex)

Ansa lenticularis?

Subthalamic nucleus

Cerebral peduncle

Forel's fields

Ventral striatum

Central nucleus

Stria terminalis

Intercalated cell groups

Basal accessory nucleus

Ansa lenticularis?

Medial forebrain bundle

Fornix

Dorsomedial nucleus

HYPO-THALAMUS

Internal capsule (posterior limb)

TEMPORAL LOBE

Anterior commissure?

Basal nucleus

Medial nucleus

Optic tract

Lateral hypothalamic area

Ventromedial nucleus

Hypothalamic G/EP

AMYGDALA

i

i

Lateral nucleus

Cortical nucleus

Lateral olfactory tract

Perifascicular GEP (surrounds optic and lateral olfactory tracts)

Cortical nucleus

Entorhinal cortex

Layer VII (subplate)

Cortical plate

Layer I

Subpial granular layer (GEP)

G/EP - glioepithelium/ependyma
GEP - glioepithelium
NEP - neuroepithelium
SVZ - subventricular zone

Germinal and transitional structures in *italics*

Lateral migratory stream (penetrates basolateral complex of amygdala)

PLATE 55A
CR 150 mm, GW 17, Y15-63
Frontal
Section 541

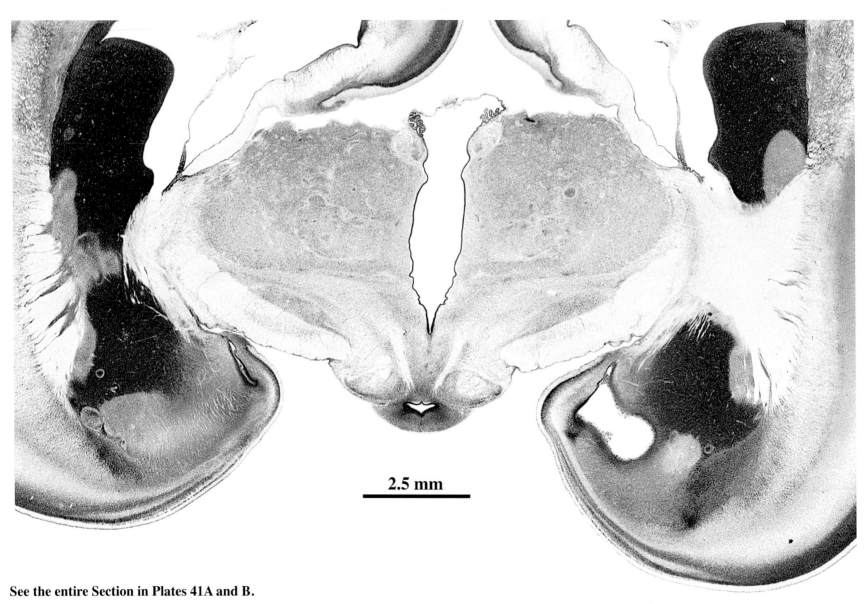

2.5 mm

See the entire Section in Plates 41A and B.

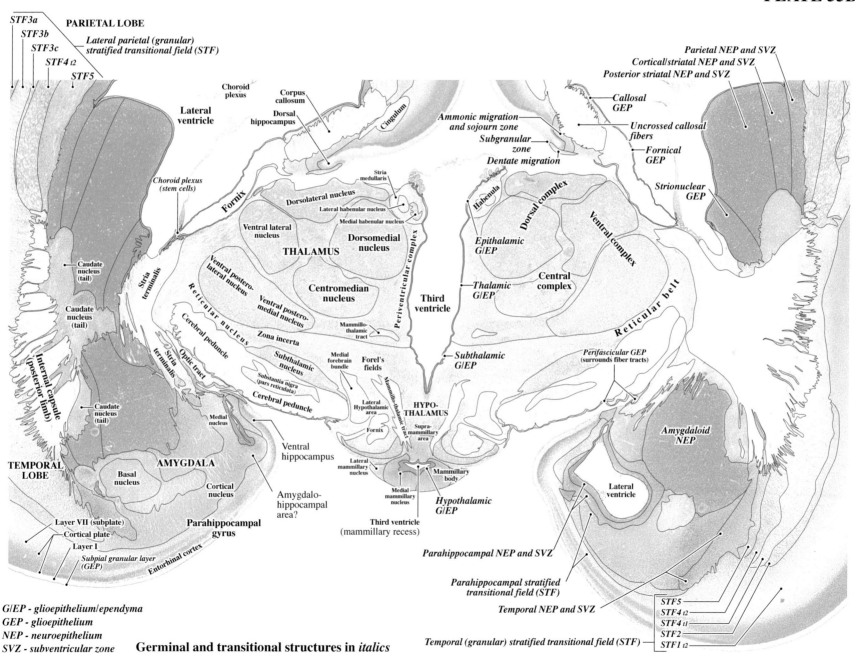

STF3a
STF3b
STF3c
STF4 t2
STF5

PARIETAL LOBE

Lateral parietal (granular)
stratified transitional field (STF)

Choroid
plexus

Corpus
callosum

Dorsal
hippocampus

**Lateral
ventricle**

*Choroid plexus
(stem cells)*

Fornix

Stria
medullaris

Dorsolateral nucleus

Lateral habenular nucleus

Medial habenular nucleus

Cingulum

Ammonic migration
and sojourn zone

Subgranular
zone

Dentate migration

Callosal
GEP

Parietal NEP and SVZ
Cortical/striatal NEP and SVZ
Posterior striatal NEP and SVZ

*Uncrossed callosal
fibers*

*Fornical
GEP*

*Strionuclear
GEP*

*Caudate
nucleus
(tail)*

*Stria
terminalis*

Ventral lateral
nucleus

THALAMUS

**Dorsomedial
nucleus**

Periventricular complex

Habenula

Dorsal complex

Ventral complex

*Epithalamic
G/EP*

*Thalamic
G/EP*

**Central
complex**

*Caudate
nucleus
(tail)*

Reticular nucleus

*Ventral postero-
lateral nucleus*

**Centromedian
nucleus**

Reticular belt

*Internal capsule
(posterior limb)*

Cerebral peduncle

Ventral postero-
medial nucleus

Zona incerta

Mammillo-
thalamic
tract

**Third
ventricle**

*Subthalamic
G/EP*

*Perifascicular GEP
(surrounds fiber tracts)*

*Caudate
nucleus
(tail)*

*Stria
terminalis*

Optic tract

Subthalamic
nucleus

*Substantia nigra
(pars reticulata)*

Medial
forebrain
bundle

Forel's
fields

Cerebral peduncle

Lateral
Hypothalamic
area

Mammillothalamic tract

**HYPO-
THALAMUS**

*Amygdaloid
NEP*

**TEMPORAL
LOBE**

*Medial
nucleus*

AMYGDALA

**Basal
nucleus**

**Cortical
nucleus**

Ventral
hippocampus

Amygdalo-
hippocampal
area?

Fornix

Supra-
mammillary
area

Lateral
ventricle

**Parahippocampal
gyrus**

Layer VII (subplate)

Cortical plate

Layer I

*Subpial granular layer
(GEP)*

Entorhinal cortex

Lateral
mammillary
nucleus

Medial
mammillary
nucleus

Mammillary
body

Third ventricle
(mammillary recess)

*Hypothalamic
G/EP*

Parahippocampal NEP and SVZ

*Parahippocampal stratified
transitional field (STF)*

Temporal NEP and SVZ

Temporal (granular) stratified transitional field (STF)

STF5
STF4 t2
STF4 t1
STF2
STF1 t2

G/EP - glioepithelium/ependyma
GEP - glioepithelium
NEP - neuroepithelium
SVZ - subventricular zone

Germinal and transitional structures in *italics*

PLATE 56A
CR 150 mm, GW 17, Y15-63
Frontal, Section 621

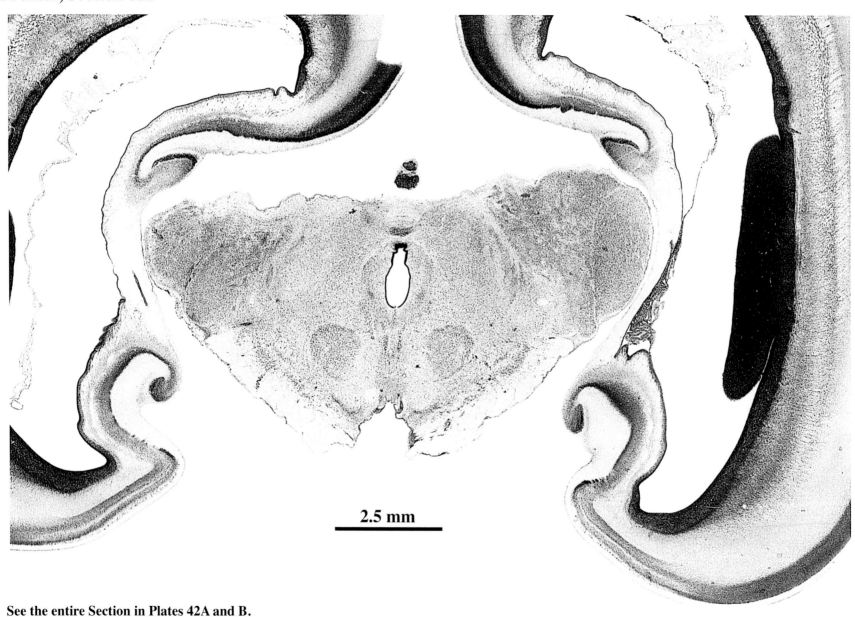

2.5 mm

See the entire Section in Plates 42A and B.

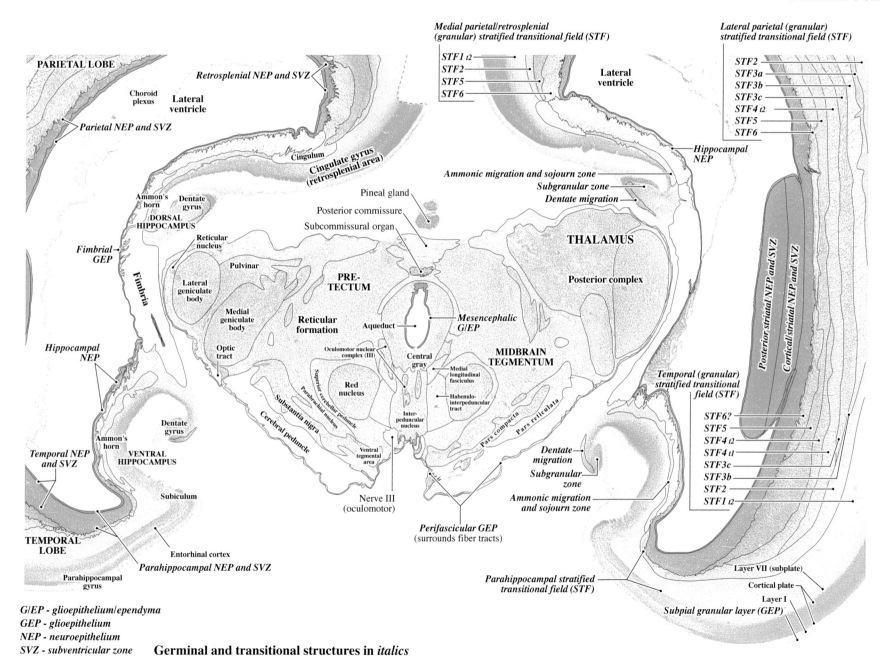

Medial parietal/retrosplenial
(granular) stratified transitional field (STF)

STF1 *t2*
STF2
STF5
STF6

Lateral parietal (granular)
stratified transitional field (STF)

STF2
STF3a
STF3b
STF3c
STF4 *t2*
STF5
STF6

PARIETAL LOBE

Retrosplenial NEP and SVZ

Choroid plexus   **Lateral ventricle**

Parietal NEP and SVZ

Cingulum

Cingulate gyrus (retrosplenial area)

Lateral ventricle

Hippocampal NEP

Ammonic migration and sojourn zone

Subgranular zone

Dentate migration

Ammon's horn   Dentate gyrus

DORSAL HIPPOCAMPUS

Pineal gland
Posterior commissure
Subcommissural organ

THALAMUS

Fimbrial GEP

Reticular nucleus

Pulvinar

Posterior complex

Fimbria

Lateral geniculate body

Medial geniculate body

PRE-TECTUM

Reticular formation

Aqueduct

Mesencephalic G/EP

Posterior striatal NEP and SVZ

Cortical striatal NEP and SVZ

Hippocampal NEP

Optic tract

Oculomotor nuclear complex (III)

Central gray

MIDBRAIN TEGMENTUM

Temporal (granular) stratified transitional field (STF)

Superior cerebellar peduncle
Parabrachial nucleus

Red nucleus

Medial longitudinal fasciculus

Habenulo-interpeduncular tract

STF6?
STF5
STF4 *t2*
STF4 *t1*
STF3c
STF3b
STF2
STF1 *t2*

Temporal NEP and SVZ

Ammon's horn   Dentate gyrus

VENTRAL HIPPOCAMPUS

Substantia nigra
Cerebral peduncle

Inter-peduncular nucleus

Pars compacta   Pars reticulata

Dentate migration

Subgranular zone

Subiculum

Ventral tegmental area

Ammonic migration and sojourn zone

TEMPORAL LOBE

Nerve III (oculomotor)

Entorhinal cortex

Parahippocampal NEP and SVZ

Perifascicular GEP
(surrounds fiber tracts)

Parahippocampal gyrus

Parahippocampal stratified transitional field (STF)

Layer VII (subplate)
Cortical plate
Layer I
Subpial granular layer (GEP)

G/EP - glioepithelium/ependyma
GEP - glioepithelium
NEP - neuroepithelium
SVZ - subventricular zone

Germinal and transitional structures in *italics*

PLATE 57A
CR 150 mm, GW 17
Y15-63
Frontal
Section 681

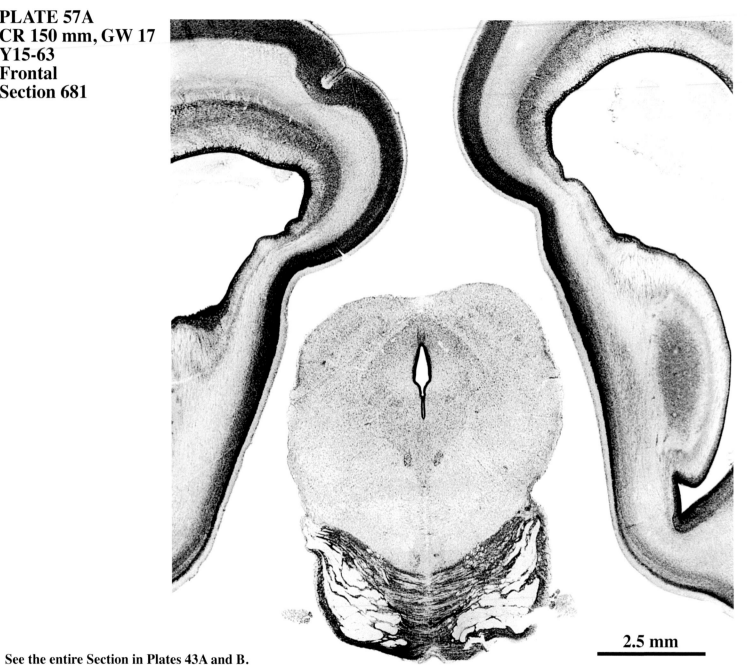

2.5 mm

See the entire Section in Plates 43A and B.

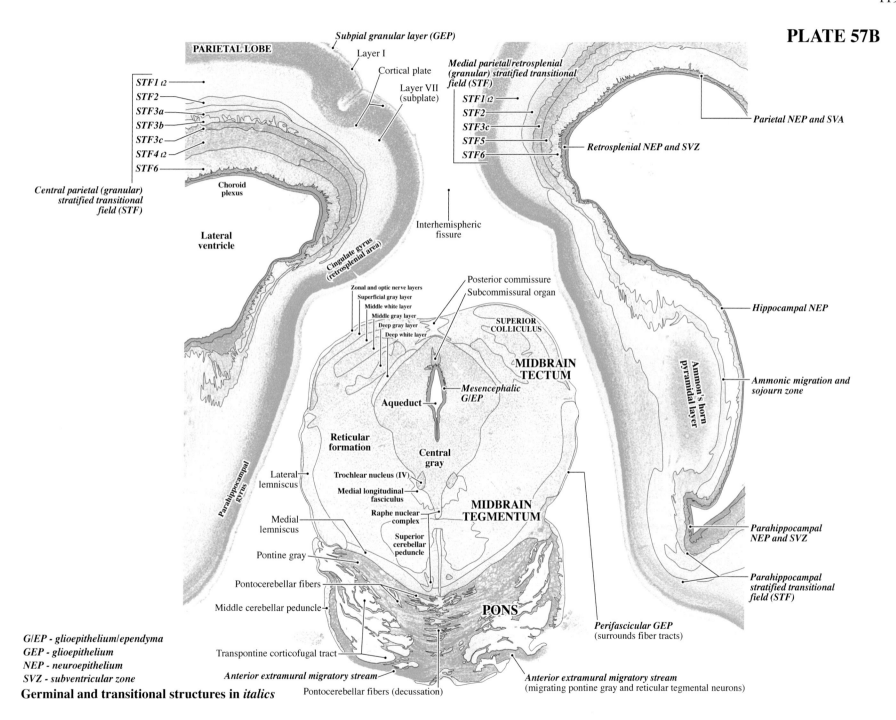

Subpial granular layer (GEP)

PARIETAL LOBE

Layer I

Cortical plate

Layer VII
(subplate)

*Medial parietal/retrosplenial
(granular) stratified transitional
field (STF)*

STF1 t2
STF2
STF3a
STF3b
STF3c
STF4 t2
STF6

STF1 t2
STF2
STF3c
STF5
STF6

Parietal NEP and SVA

Retrosplenial NEP and SVZ

*Central parietal (granular)
stratified transitional
field (STF)*

Choroid
plexus

**Lateral
ventricle**

Interhemispheric
fissure

Cingulate gyrus
(retrosplenial area)

Hippocampal NEP

Posterior commissure
Subcommissural organ

Zonal and optic nerve layers
Superficial gray layer
Middle white layer
Middle gray layer
Deep gray layer
Deep white layer

**SUPERIOR
COLLICULUS**

**MIDBRAIN
TECTUM**

*Ammonic migration and
sojourn zone*

Ammon's horn
pyramidal layer

*Mesencephalic
G/EP*

Aqueduct

**Reticular
formation**

**Central
gray**

Lateral
lemniscus

Trochlear nucleus (IV)

Medial longitudinal
fasciculus

Raphe nuclear
complex

**MIDBRAIN
TEGMENTUM**

*Parahippocampal
NEP and SVZ*

Parahippocampal gyrus

Medial
lemniscus

Superior
cerebellar
peduncle

Pontine gray

*Parahippocampal
stratified transitional
field (STF)*

Pontocerebellar fibers

Middle cerebellar peduncle

PONS

Perifascicular GEP
(surrounds fiber tracts)

G/EP - glioepithelium/ependyma
GEP - glioepithelium
NEP - neuroepithelium
SVZ - subventricular zone

Transpontine corticofugal tract

Anterior extramural migratory stream

Pontocerebellar fibers (decussation)

Anterior extramural migratory stream
(migrating pontine gray and reticular tegmental neurons)

Germinal and transitional structures in *italics*

PLATE 58A
CR 150 mm
GW 17
Y15-63
Frontal
Section 741

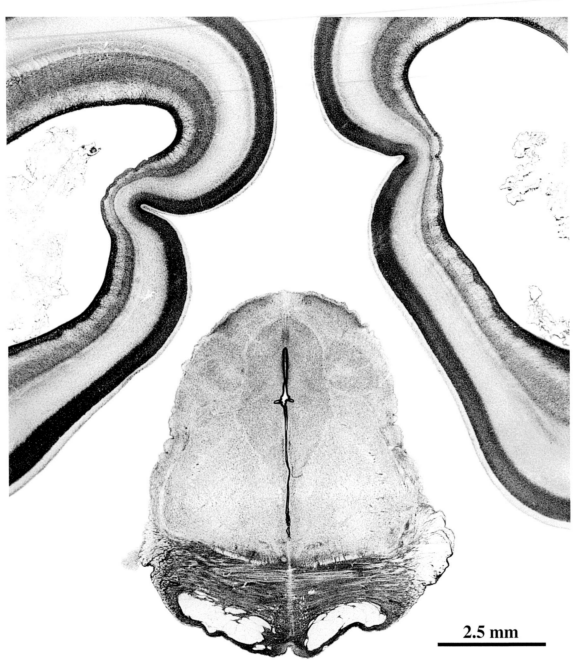

2.5 mm

See the entire Section in Plates 44A and B.

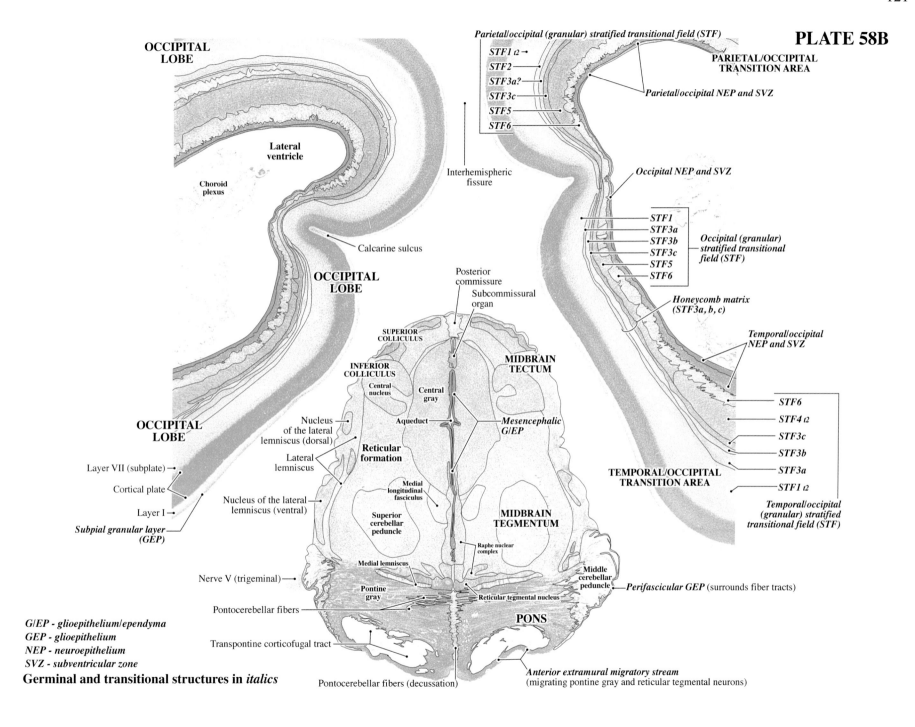

PLATE 58B

OCCIPITAL LOBE

Lateral ventricle

Choroid plexus

Calcarine sulcus

OCCIPITAL LOBE

OCCIPITAL LOBE

Layer VII (subplate)

Cortical plate

Layer I

Subpial granular layer (GEP)

Nerve V (trigeminal)

Pontocerebellar fibers

Transpontine corticofugal tract

Pontocerebellar fibers (decussation)

Parietal/occipital (granular) stratified transitional field (STF)

STF1 t2 →
STF2
STF3a?
STF3c
STF5
STF6

PARIETAL/OCCIPITAL TRANSITION AREA

Parietal/occipital NEP and SVZ

Interhemispheric fissure

Occipital NEP and SVZ

STF1
STF3a
STF3b
STF3c
STF5
STF6

Occipital (granular) stratified transitional field (STF)

Honeycomb matrix (STF3a, b, c)

Temporal/occipital NEP and SVZ

Posterior commissure
Subcommissural organ

SUPERIOR COLLICULUS

INFERIOR COLLICULUS

Central nucleus

Central gray

Aqueduct

Mesencephalic G/EP

MIDBRAIN TECTUM

Nucleus of the lateral lemniscus (dorsal)

Reticular formation

Lateral lemniscus

Medial longitudinal fasciculus

Nucleus of the lateral lemniscus (ventral)

Superior cerebellar peduncle

MIDBRAIN TEGMENTUM

Raphe nuclear complex

Medial lemniscus

Pontine gray

Reticular tegmental nucleus

Middle cerebellar peduncle

Perifascicular GEP (surrounds fiber tracts)

PONS

STF6
STF4 t2
STF3c
STF3b
STF3a
STF1 t2

TEMPORAL/OCCIPITAL TRANSITION AREA

Temporal/occipital (granular) stratified transitional field (STF)

G/EP - glioepithelium/ependyma
GEP - glioepithelium
NEP - neuroepithelium
SVZ - subventricular zone

Germinal and transitional structures in *italics*

Anterior extramural migratory stream
(migrating pontine gray and reticular tegmental neurons)

PLATE 59A
CR 150 mm, GW 17, Y15-63
Frontal
Section 801

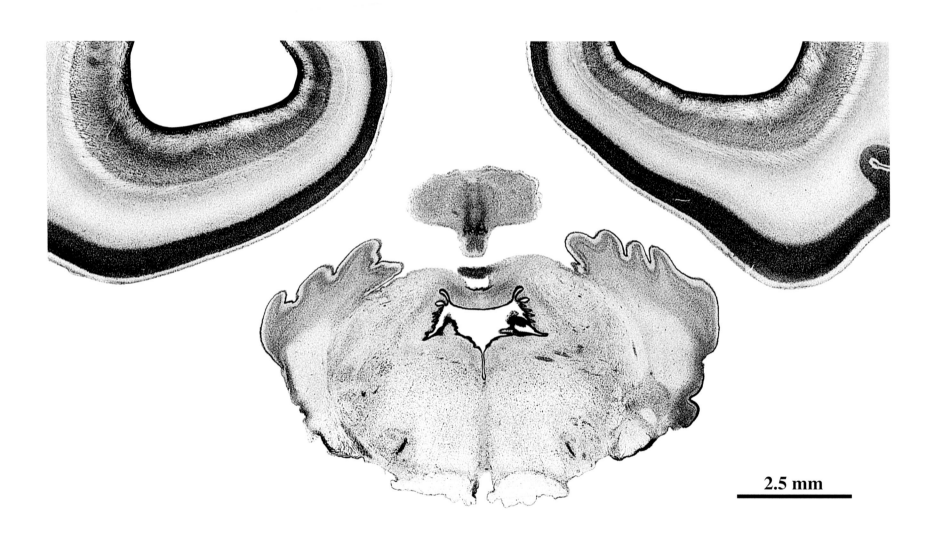

2.5 mm

See the entire Section in Plates 45A and B.

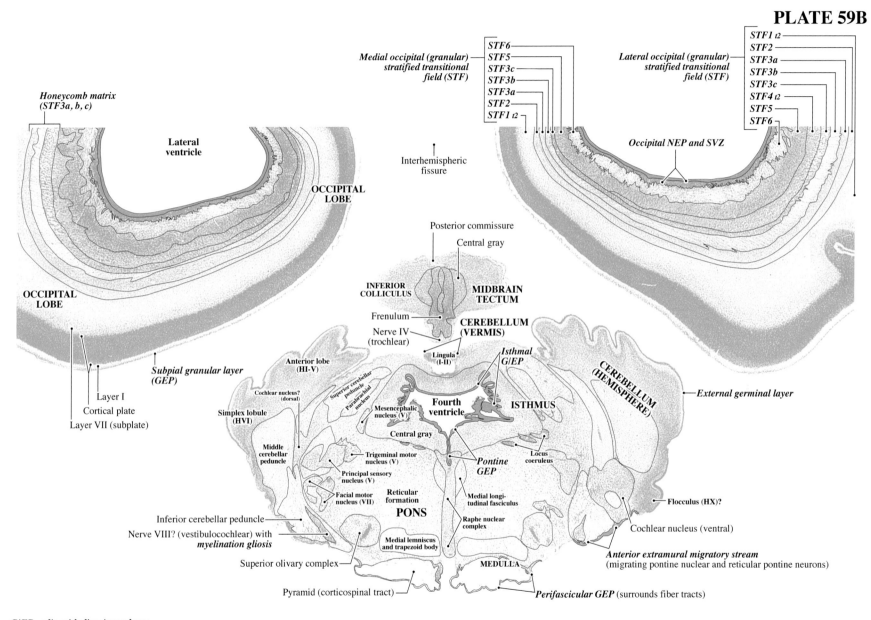

Honeycomb matrix
(STF3a, b, c)

Lateral
ventricle

**OCCIPITAL
LOBE**

**OCCIPITAL
LOBE**

STF6
STF5
STF3c
STF3b
STF3a
STF2
STF1 t2

Medial occipital (granular)
stratified transitional
field (STF)

STF1 t2
STF2
STF3a
STF3b
STF3c
STF4 t2
STF5
STF6

Lateral occipital (granular)
stratified transitional
field (STF)

Occipital NEP and SVZ

Interhemispheric
fissure

Posterior commissure
Central gray

**INFERIOR
COLLICULUS**

**MIDBRAIN
TECTUM**

Frenulum

Nerve IV
(trochlear)

**CEREBELLUM
(VERMIS)**

Lingula
(I-II)

*Isthmal
G/EP*

**CEREBELLUM
(HEMISPHERE)**

Anterior lobe
(HI-V)

Cochlear nucleus?
(dorsal)

Superior cerebellar
peduncle

Parabrachial
nucleus

Mesencephalic
nucleus (V)

**Fourth
ventricle**

ISTHMUS

External germinal layer

Subpial granular layer
(GEP)

Layer I

Cortical plate

Layer VII (subplate)

Simplex lobule
(HVI)

Middle
cerebellar
peduncle

Trigeminal motor
nucleus (V)

Principal sensory
nucleus (V)

Facial motor
nucleus (VII)

Central gray

*Pontine
GEP*

Locus
coeruleus

**Reticular
formation**

PONS

Medial longi-
tudinal fasciculus

Raphe nuclear
complex

Flocculus (HX)?

Cochlear nucleus (ventral)

Inferior cerebellar peduncle

Nerve VIII? (vestibulocochlear) with
myelination gliosis

Superior olivary complex

Medial lemniscus
and trapezoid body

MEDULLA

Anterior extramural migratory stream
(migrating pontine nuclear and reticular pontine neurons)

Pyramid (corticospinal tract)

Perifascicular GEP (surrounds fiber tracts)

G/EP - glioepithelium/ependyma
GEP - glioepithelium
NEP - neuroepithelium
SVZ - subventricular zone
Germinal and transitional structures in *italics*

PLATE 60A
CR 150 mm, GW 17, Y15-63
Frontal
Section 821

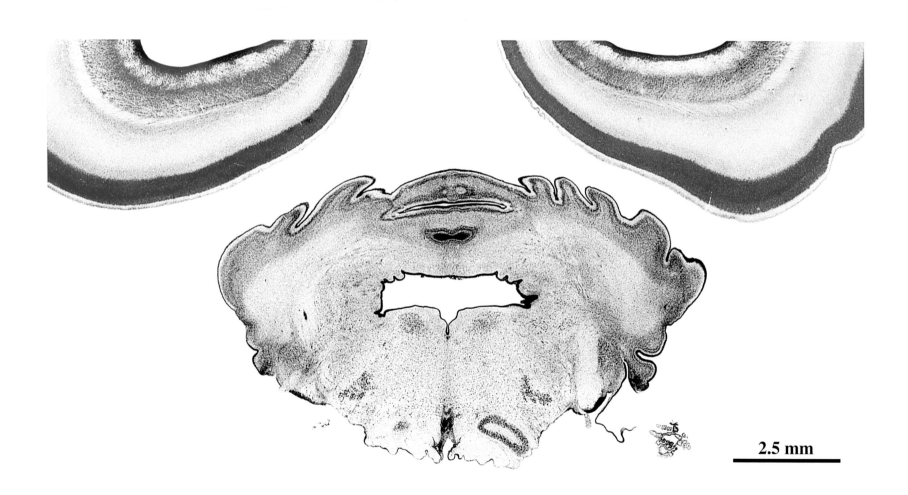

2.5 mm

See the entire Section in Plates 46A and B.

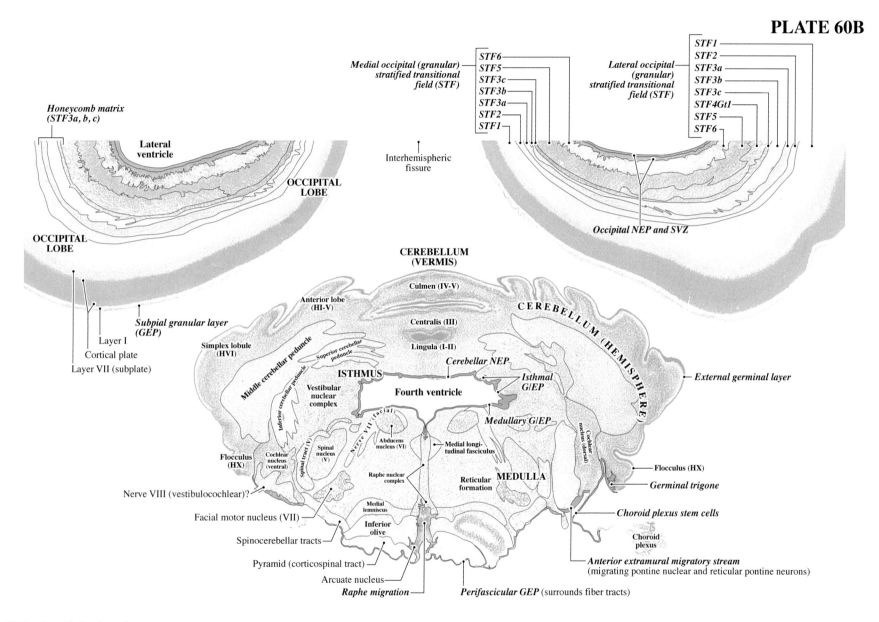

Honeycomb matrix (STF3a, b, c)

Medial occipital (granular) stratified transitional field (STF)
STF6
STF5
STF3c
STF3b
STF3a
STF2
STF1

Lateral occipital (granular) stratified transitional field (STF)
STF1
STF2
STF3a
STF3b
STF3c
STF4Gtl
STF5
STF6

Lateral ventricle

OCCIPITAL LOBE

Interhemispheric fissure

Occipital NEP and SVZ

OCCIPITAL LOBE

Subpial granular layer (GEP)

Layer I
Cortical plate
Layer VII (subplate)

CEREBELLUM (VERMIS)

Culmen (IV-V)

Anterior lobe (HI-V)

Centralis (III)

Lingula (I-II)

C E R E B E L L U M (H E M I S P H E R E)

Simplex lobule (HVI)

Middle cerebellar peduncle
Superior cerebellar peduncle

Cerebellar NEP

Isthmal G/EP

External germinal layer

ISTHMUS

Inferior cerebellar peduncle

Vestibular nuclear complex

Fourth ventricle

Medullary G/EP

Nerve VII (facial)

Abducens nucleus (VI)

Spinal tract (V)

Spinal nucleus (V)

Medial longitudinal fasciculus

Cochlear nucleus (dorsal)

Flocculus (HX)

Cochlear nucleus (ventral)

Raphe nuclear complex

Reticular formation

MEDULLA

Flocculus (HX)

Germinal trigone

Nerve VIII (vestibulocochlear)?

Medial lemniscus

Choroid plexus stem cells

Facial motor nucleus (VII)

Inferior olive

Choroid plexus

Spinocerebellar tracts

Anterior extramural migratory stream (migrating pontine nuclear and reticular pontine neurons)

Pyramid (corticospinal tract)

Arcuate nucleus

Raphe migration

Perifascicular GEP (surrounds fiber tracts)

G/EP - glioepithelium/ependyma
GEP - glioepithelium
NEP - neuroepithelium
SVZ - subventricular zone

Germinal and transitional structures in *italics*

PLATE 61A
CR 150 mm, GW 17, Y15-63
Frontal
Section 861

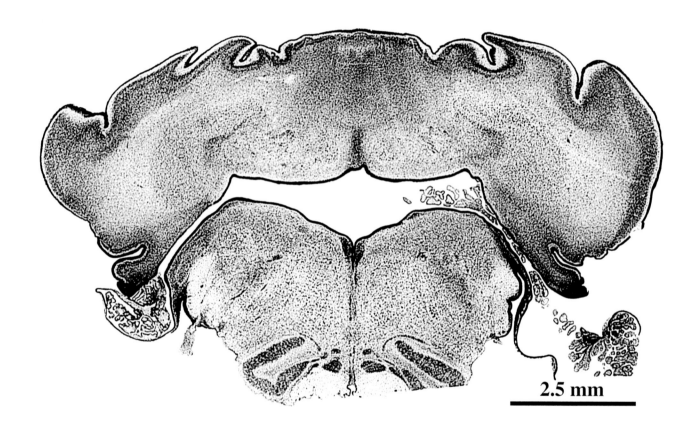

2.5 mm

See the entire Section in Plates 47A and B.

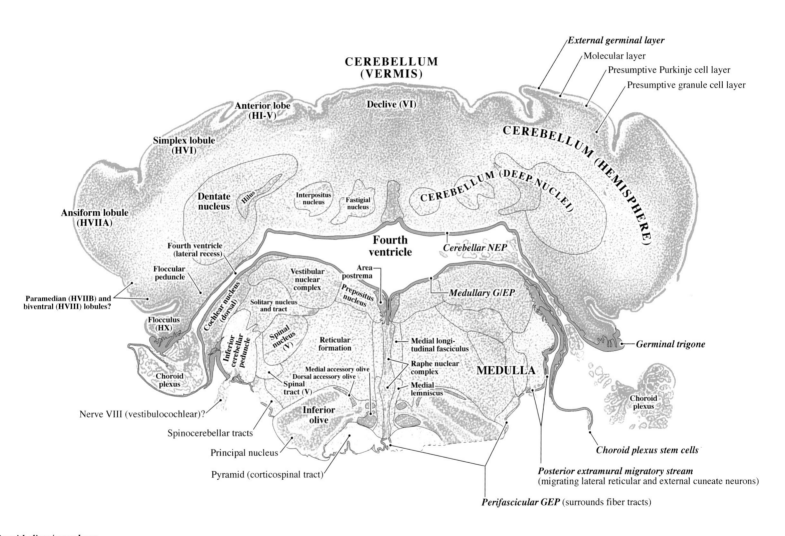

**CEREBELLUM
(VERMIS)**

External germinal layer
Molecular layer
Presumptive Purkinje cell layer
Presumptive granule cell layer

Anterior lobe
(HI-V)

Declive (VI)

Simplex lobule
(HVI)

CEREBELLUM (HEMISPHERE)

Dentate
nucleus

Hilus

Interpositus
nucleus

Fastigial
nucleus

CEREBELLUM (DEEP NUCLEI)

Ansiform lobule
(HVIIA)

Fourth ventricle
(lateral recess)

**Fourth
ventricle**

Cerebellar NEP

Floccular
peduncle

Vestibular
nuclear
complex

Area
postrema

Cochlear nucleus (dorsal)

Prepositus
nucleus

Medullary G/EP

Paramedian (HVIIB) and
biventral (HVIII) lobules?

Solitary nucleus
and tract

Flocculus
(HX)

Inferior cerebellar peduncle

Spinal
nucleus
(V)

Reticular
formation

Medial longi-
tudinal fasciculus

Germinal trigone

Choroid
plexus

Medial accessory olive
Dorsal accessory olive
Spinal
tract (V)

Raphe nuclear
complex

MEDULLA

Medial
lemniscus

Nerve VIII (vestibulocochlear)?

**Inferior
olive**

*Choroid
plexus*

Spinocerebellar tracts

Choroid plexus stem cells

Principal nucleus

Pyramid (corticospinal tract)

Posterior extramural migratory stream
(migrating lateral reticular and external cuneate neurons)

Perifascicular GEP (surrounds fiber tracts)

G/EP - glioepithelium/ependyma
GEP - glioepithelium
NEP - neuroepithelium

Germinal and transitional structures in *italics*

PLATE 62A
CR 150 mm, GW 17, Y15-63
Frontal
Section 911

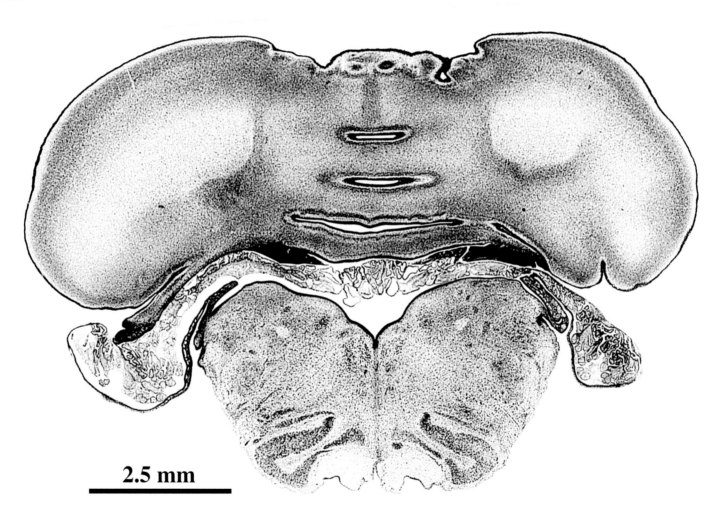

2.5 mm

See the entire Section in Plates 48A and B.

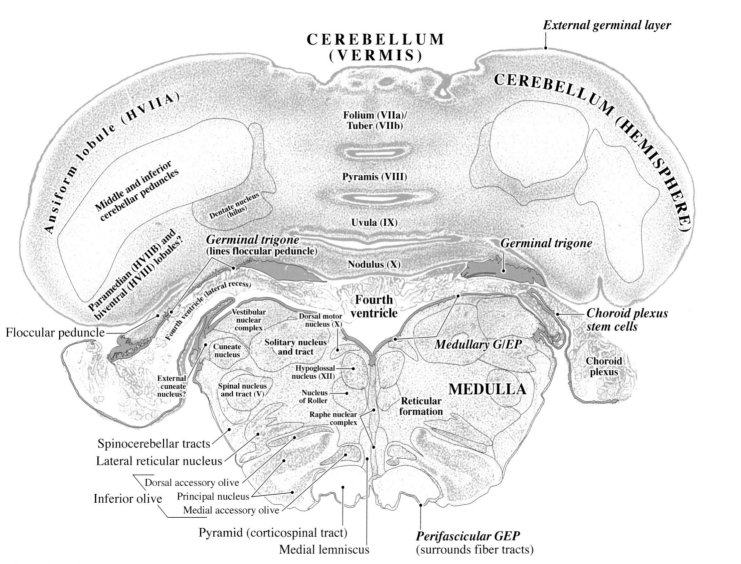

CEREBELLUM
(VERMIS)

External germinal layer

CEREBELLUM (HEMISPHERE)

Ansiform lobule (HVIIA)

Middle and inferior cerebellar peduncles

Folium (VIIa)/ Tuber (VIIb)

Pyramis (VIII)

Dentate nucleus (hilus)

Uvula (IX)

Paramedian (HVIIB) and biventral (HVIII) lobules?

Germinal trigone (lines floccular peduncle)

Nodulus (X)

Germinal trigone

Fourth ventricle

Fourth ventricle (lateral recess)

Floccular peduncle

Vestibular nuclear complex

Dorsal motor nucleus (X)

Solitary nucleus and tract

Choroid plexus stem cells

Medullary G/EP

Choroid plexus

Cuneate nucleus

External cuneate nucleus?

Hypoglossal nucleus (XII)

Spinal nucleus and tract (V)

Nucleus of Roller

MEDULLA

Raphe nuclear complex

Reticular formation

Spinocerebellar tracts

Lateral reticular nucleus

Dorsal accessory olive

Inferior olive

Principal nucleus

Medial accessory olive

Pyramid (corticospinal tract)

Medial lemniscus

Perifascicular GEP (surrounds fiber tracts)

G/EP - glioepithelium/ependyma
GEP - glioepithelium
Germinal and transitional structures in *italics*

PLATE 63
CR 150 mm, GW 17
Y15-63
Section 341
FRONTAL AGRANULAR
CORTEX

See the complete Section
in Plates 36A and B.

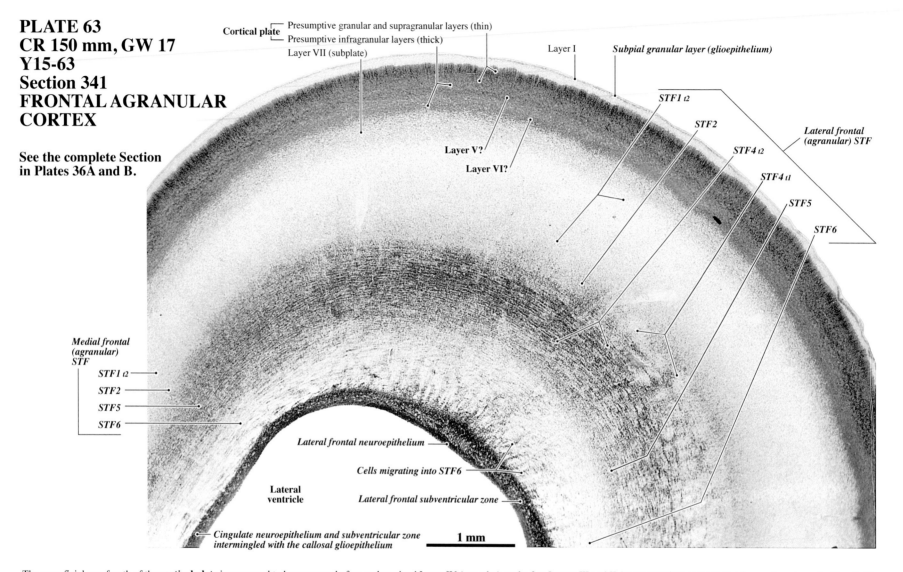

Cortical plate — Presumptive granular and supragranular layers (thin)
— Presumptive infragranular layers (thick)
Layer VII (subplate)

Layer I

Subpial granular layer (glioepithelium)

STF1 t2

STF2

Lateral frontal (agranular) STF

Layer V?

Layer VI?

STF4 t2

STF4 t1

STF5

STF6

Medial frontal (agranular) STF

STF1 t2

STF2

STF5

STF6

Lateral frontal neuroepithelium

Cells migrating into STF6

Lateral ventricle

Lateral frontal subventricular zone

Cingulate neuroepithelium and subventricular zone
intermingled with the callosal glioepithelium

1 mm

The superficial one-fourth of the **cortical plate** is presumed to be composed of recently arrived Layer IV (granular), and a few Layers III and II (supragranular) neurons that form a densely packed band. The deep three-fourths of the cortical plate is composed mainly of Layers V and VI infragranular neurons; Layer V may be the slightly denser band in the middle of the cortical plate, and Layer VI may be the less dense band at its base. Since frontal cortex does not have a prominent Layer IV, the granular and supragranular layers are thin and the infragranular layers are thick. In addition many neurons that will occupy the granular and supragranular layers are still migrating and sojourning in the *stratified transitional field (STF)*.

The thickest part of the cortex is the *STF* between Layer VII and the densely packed *neuroepithelial/subventricular* cells lining the ventricle. In this area of agranular (sparse Layer IV) frontal cortex, five of the six *STF* layers are distinguishable. *STF1 t2* is thick and sparsely cellular. *STF4* is the thickest layer with a superficial *t1* phase (fibers predominate before cells migrate through on their way to the cortical plate) and a deep *t2* phase (cells intermingle with fibers to create striations in the layer). *STF5* is indefinite and disappearing laterally, but is thick and dense in the medial frontal cortex. *STF6* is full of infiltrating callosal fibers intermingling with cells migrating out of the subventricular zone. *STF3* is absent in agranular cortical areas.

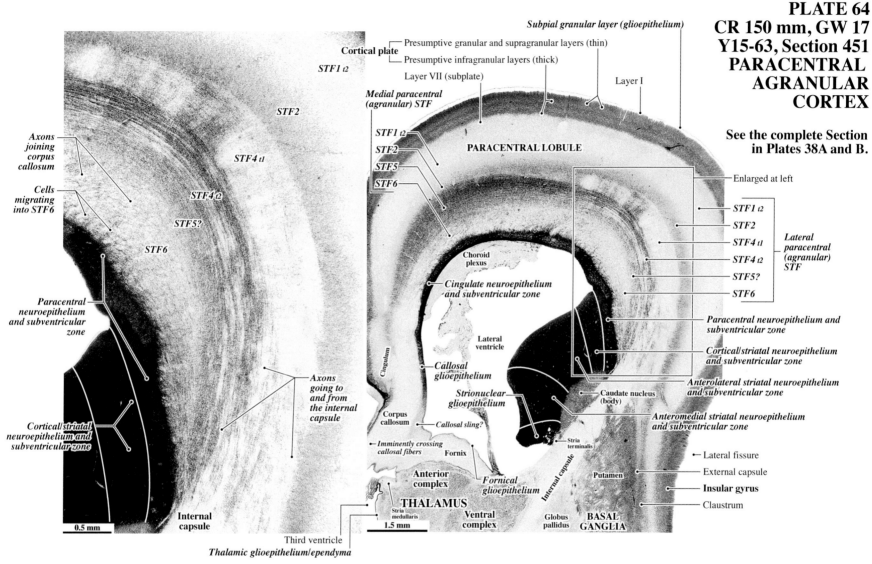

PLATE 64
CR 150 mm, GW 17
Y15-63, Section 451
PARACENTRAL
AGRANULAR
CORTEX

See the complete Section in Plates 38A and B.

The *white lines* in the **neuroepithelium and subventricular zone** are *presumptive* subdivisions. Since the striatal and cortical germinal matrices are continuous, the junction (**cortical/striatal neuroepithelium and subventricular zone**) is presumed to contain stem cells that generate neurons and glia in both structures. Lamination in the paracentral **cortical plate** in this section is similar to frontal cortex in Section 341 (**Plate 63**, facing page) with thin granular/supragranular and thick infragranular parts. That suggests that this section cuts through the agranular portion (presumptive motor area) of the paracentral lobule, where Layer IV is not prominent.

Layers in the **stratified transitional field (STF)** are more pronounced than in Section 341 (**Plate 63**) with prominent lateral-to-medial differences in **STF2**, **STF4**, and **STF5**. **STF2** is cell-dense and thick laterally, cell-sparse and thin medially. **STF4** is absent medially, but it is thick laterally and subdivided into an early-developing superficial part (**t1**) and a late-developing deep part (**t2**). Medially, **STF5** is the thickest and densest cellular layer; laterally it in infiltrated by fibers presumably going to and from the internal capsule.

PLATE 65
CR 150 mm
GW 17
Y15-63
Section 501
PARACENTRAL
GRANULAR
CORTEX

See the complete Section in Plates 40A and B.

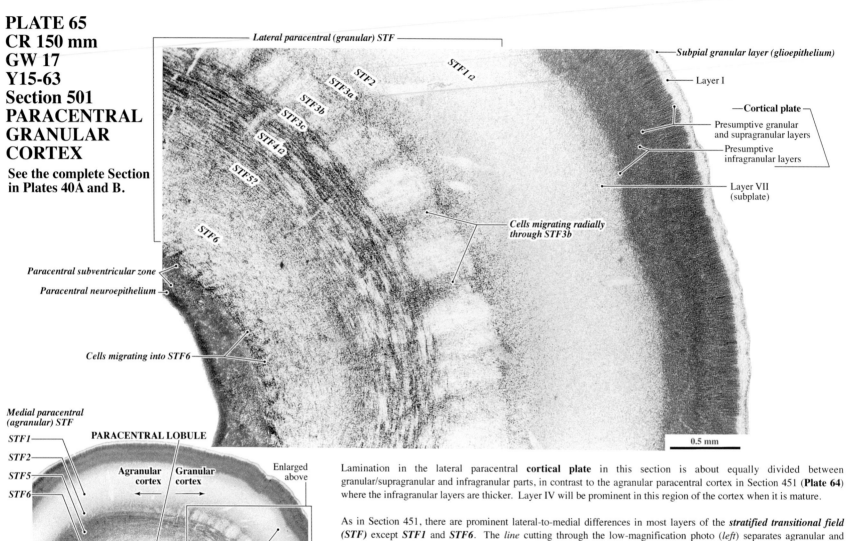

Lateral paracentral (granular) STF

Subpial granular layer (glioepithelium)

STF1 t2

STF2

STF3a

STF3b

STF3c

STF4 t2

STF5?

STF6

Layer I

Cortical plate

Presumptive granular and supragranular layers

Presumptive infragranular layers

Layer VII (subplate)

Cells migrating radially through STF3b

Paracentral subventricular zone

Paracentral neuroepithelium

Cells migrating into STF6

0.5 mm

Medial paracentral (agranular) STF

STF1

STF2

STF5

STF6

PARACENTRAL LOBULE

Agranular cortex ← → Granular cortex

Enlarged above

1 mm

Lateral paracentral (granular) STF

Lamination in the lateral paracentral **cortical plate** in this section is about equally divided between granular/supragranular and infragranular parts, in contrast to the agranular paracentral cortex in Section 451 (**Plate 64**) where the infragranular layers are thicker. Layer IV will be prominent in this region of the cortex when it is mature.

As in Section 451, there are prominent lateral-to-medial differences in most layers of the *stratified transitional field* (*STF*) except *STF1* and *STF6*. The *line* cutting through the low-magnification photo (*left*) separates agranular and granular areas of the paracentral lobule. In the medial paracentral cortex *STF2* is thin and *STF5* is thick and dense. *STF3* is completely absent, since this is agranular cortex. In the lateral paracentral cortex (*above*), *STF2* is thick but contains only sparsely scattered cells. *STF3a* is the thin and dense cellular layer at the base of *STF2*. *STF3b* is a narrow middle fibrous layer that is postulated to contain mainly cortical afferent axons; streaks of radially arrayed cell groups appear to be migrating through the fibers. *STF3c* is a thin, dense cellular layer inferior to the fibers. *STF4 t2* is postulated to contain many cells pausing in their migration to mingle with afferent axons entering the cortex from the internal capsule, creating dark, horizontal streaks among the fibers. The least prominent layer in the lateral paracentral cortex is *STF5* that is presumed to contain younger cells that have not yet been invaded by afferent axons.

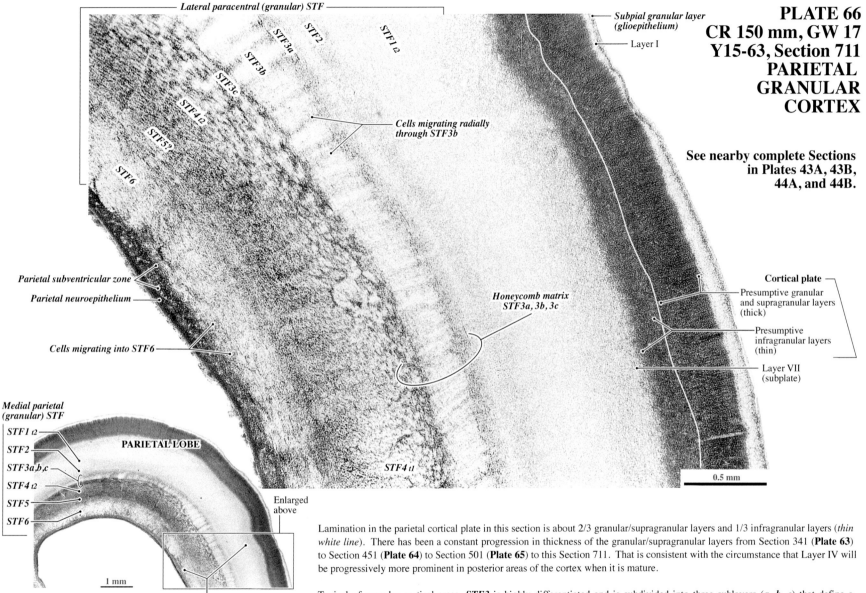

Lateral paracentral (granular) STF

STF2
STF3a
STF3b
STF3c
STF4 t2
STF5?
STF6
STF1 t2

Subpial granular layer (glioepithelium)
Layer I

STF1 t2

Cells migrating radially through STF3b

PLATE 66
CR 150 mm, GW 17
Y15-63, Section 711
PARIETAL
GRANULAR
CORTEX

See nearby complete Sections in Plates 43A, 43B, 44A, and 44B.

Parietal subventricular zone

Parietal neuroepithelium

Cells migrating into STF6

Honeycomb matrix
STF3a, 3b, 3c

Cortical plate
Presumptive granular and supragranular layers (thick)

Presumptive infragranular layers (thin)

Layer VII (subplate)

STF4 t1

0.5 mm

Medial parietal (granular) STF

STF1 t2
STF2
STF3a,b,c
STF4 t2
STF5
STF6

PARIETAL LOBE

Enlarged above

1 mm

Lateral parietal (granular) STF

Lamination in the parietal cortical plate in this section is about 2/3 granular/supragranular layers and 1/3 infragranular layers (*thin white line*). There has been a constant progression in thickness of the granular/supragranular layers from Section 341 (**Plate 63**) to Section 451 (**Plate 64**) to Section 501 (**Plate 65**) to this Section 711. That is consistent with the circumstance that Layer IV will be progressively more prominent in posterior areas of the cortex when it is mature.

Typical of granular cortical areas, **STF3** is highly differentiated and is subdivided into three sublayers (*a, b, c*) that define a **honeycomb matrix**. **STF3a** is the thin and dense cellular layer at the base of **STF2**. **STF3b** is a narrow middle fibrous layer that is postulated to contain mainly cortical afferent axons; streaks of radially arrayed cell groups appear to be migrating through the fibers; it is these cells that define honeycomb-like chambers among the fibers. **STF3c** is a thin, dense cellular layer inferior to the **STF3b** fibers.